Introduction

The aim of the *Primary Mathematics* curriculum is to allow students t[illegible]ility in mathematical problem solving. This includes using and applying mathema[illegible]ctical, real-life situations as well as within the discipline of mathematics itself. Therefore the curriculum covers a wide range of situations from routine problems to problems in unfamiliar contexts to open-ended investigations that make use of relevant mathematical concepts.

An important feature of learning mathematics with this curriculum is the use of a concrete introduction to the concept, followed by a pictorial representation, followed by the abstract symbols. The textbook supplies the pictorial and abstract aspects of this progression. You, as the teacher, should supply the concrete introduction. For some students a concrete illustration is more important than for other students.

This guide includes the following :

- **Scheme of Work**: A table with a suggested weekly schedule, the primary objective for each lesson, and corresponding pages from the textbook, workbook, and guide.
- **Manipulatives**: A list of manipulatives used in this guide.
- **Objectives**: A list of objectives for each chapter.
- **Vocabulary**: A list of new mathematical terms.
- **Notes**: An explanation of what students should know before starting the chapter, the concepts that will be covered in the chapter, and how these concepts fit in with the program as a whole.
- **Material**: A list of suggested manipulatives that can be used in presenting the concepts in each chapter.
- **Activity**: Teaching activities to introduce a concept concretely or to follow up on a concept in order to clarify or extend it so that students will be more successful with independent practice.
- **Discussion**: The opening pages of the chapter and tasks in the textbook that should be discussed with the student. A scripted discussion is not provided. You should follow the material in the textbook. Additional pertinent points that should be included in the discussion are given in this guide.
- **Practice**: Tasks in the textbooks students can do as guided practice or as an assessment to see if they understood the concepts covered in the teaching activity or the discussions.
- **Workbook**: Workbook exercise that should be done after the lesson.
- **Reinforcement**: Additional activities that can be used if your student needs more practice or reinforcement of the concepts. This includes references to the exercises in the optional *Primary Mathematics Extra Practice* book.
- **Games**: Optional simple games that can be used to practice skills.

- **Enrichment**: Optional activities that can be used to further explore the concepts or to provide some extra challenge.
- **Tests:** References to the appropriate tests in the *Primary Mathematics Tests* book.
- **Answers:** Answers to all the textbook tasks and workbook problems, and many fully worked solutions. Answers to textbook tasks are provided within the lesson. Answers to workbook exercises for the chapter are located at the end of each chapter in the guide.
- **Mental Math**: Problems for more practice with mental math strategies.
- **Appendix**: Pages containing drawings and charts that can be copied and used in the lessons.

The textbook and workbook both contain a review for every unit. You can use these in any way beneficial to your student. For students who benefit from a more continuous review, you can assign three problems or so a day from one of the practices or reviews. Or, you can use the reviews to assess any misunderstanding before administering a test. The reviews, particularly in the textbook, do sometimes carry the concepts a little farther. They are cumulative, and so allow you to refresh your student's memory or understanding on a topic that was covered earlier in the year.

In addition, there are supplemental books for *Extra Practice* and *Tests*. In the test book, there are two tests for each section. The second test is multiple choice. There is also a set of two cumulative tests at the end of each unit. You do not need to use both tests. If you use only one test, you can save the other for review or practice later on. You can even use the review in the workbook as a test and not get the test book at all. So there are plenty of choices for assessment, review, and practice.

The mental math exercises that go along with a particular chapter or lesson are listed as reinforcement in the lesson. They can be used in a variety of ways. You do not need to use all the mental math exercises listed for a lesson on the day of the lesson. In some review lessons, they can be used for the independent work, since there is not always a workbook exercise. You can have your student do one mental math exercise a day, repeating some of them, at the start of the lesson or as part of the independent work. You can do them orally, or have your student fill in the blanks. You can have your student do a 1-minute "sprint" at the start of each lesson using one mental math exercise for several days to see if he or she can get more of the problems done each successive day. You can use the mental math exercises as a guide for creating additional "drill" exercises.

This "Scheme of Work" on the next few pages is a suggested weekly schedule to help you keep on track for finishing the textbook in about 18 weeks. No one schedule or curriculum can meet the needs of all students equally. For some chapters, your student may be able to do the work more quickly, and for others more slowly. Take the time your student needs on each topic and each lesson. For students with a good mathematical background, each lesson in this guide will probably take a day. For others, some lessons which include a review of previously covered concepts may take more than a day. There are also optional lessons that are entirely review and a few optional mental math lessons.

Use the reinforcement or extension activities at your discretion. You may want to include extra days to play some of the games or to allow time for more practice, such as in memorizing the multiplication and division facts.

Scheme of Work

Textbook: *Primary Mathematics Textbook* 3B, Standards Edition
Workbook: *Primary Mathematics Workbook* 3B, Standards Edition
Guide: *Primary Mathematics Home Instructor's Guide* 3B, Standards Edition (this book)
Extra Practice: *Primary Mathematics Extra Practice* 3, Standards Edition
Tests: *Primary Mathematics Tests* 3B, Standards Edition

Week		Objectives	Text book	Work book	Guide
Unit 6: Length					
	Chapter 1: Meters and Centimeters				1-2
1	1	Measure in meters and centimeters	8-10	7	3-4
	2	Subtract from meters	11	8	5
	3	Add and subtract meters and centimeters	12	9-10	6-7
		Extra Practice, Unit 6, Exercise 1, pp. 113-114			
		Tests, Unit 6, 1A and 1B, pp. 1-6			
	Chapter 2: Kilometers				8
	1	Measure in kilometers and meters	13-15	11-14	9
	2	Subtract from kilometers	16-17	15-16	10
2	3	Add and subtract kilometers and meters	18	17-18	11-12
		Extra Practice, Unit 6, Exercise 2, pp. 115-116			
	4	Practice	19		13
		Tests, Unit 6, 2A and 2B, pp. 7-12			
	Chapter 3: Yards, Feet and Inches				14
	1	Convert between yards and feet	20-21		15-16
	2	Convert between feet and inches	21-22	19-20	17
	3	Add and subtract yards and feet in compound units	22-23		18
3	4	Add and subtract feet and inches in compound units	23-24	21-22	19
	5	Practice	25		20
		Extra Practice, Unit 6, Exercise 3, pp. 117-118			
		Tests, Unit 6, 3A and 3B, pp. 13-20			
	Chapter 4: Miles				21
	1	Understand miles as a unit of measurement	26	23	22
		Extra Practice, Unit 6, Exercise 4, pp. 119-120			
		Tests, Unit 6, 4A and 4B, pp. 21-28			
	Review 6		27-29	24-27	23
		Tests, Units 1-6, Cumulative A and B, pp. 29-40			
	Answers for unit 6 workbook exercises and review				24-26

Week		Objectives	Text book	Work book	Guide
Unit 7: Weight					
	Chapter 1: Kilograms and Grams				27
4	1	Weigh objects in kilograms and grams	30-31	28-29	28
	2	Convert between kilograms and grams	32	30-33	29
	3	Add and subtract kilograms and grams	33-35	34-36	30
		Extra Practice, Unit 7, Exercise 1, pp. 127-130			
		Tests, Unit 7, 1A and 1B, pp. 41-47			
	Chapter 2: Word Problems				31-32
	1	Solve word problems	36-39	37-39	33-34
		Extra Practice, Unit 7, Exercise 2, pp. 131-132			
	2	Practice	40		35
		Tests, Unit 7, 2A and 2B, pp. 49-55			
	Chapter 3: Pounds and Ounces				36
5	1	Convert between pounds and ounces	41-43	40-41	37-38
	2	Add and subtract in pounds and ounces	44-45	42-43	39
		Extra Practice, Unit 7, Exercise 3, pp. 133-134			
		Tests, Unit 7, 3A and 3B, pp. 57-63			
	Review 7		46-47	44-48	40
		Tests, Units 1-7, Cumulative A and B, pp. 65-76			
	Answers for unit 7 workbook exercises and review				41-43
Unit 8: Capacity					
	Chapter 1: Liters and Milliliters				44
	1	Measure in liters and milliliters	48-52	49-50	45
6	2	Convert between liters and milliliters	53-54	51-53	46
	3	Add and subtract liters and milliliters	55-56	54-58	47
	4	Practice	57		48
		Extra Practice, Unit 8, Exercise 1, pp. 139-142			
		Tests, Unit 8, 1A and 1B, pp. 77-83			
	Chapter 2: Gallons, Quarts, Pints and Cups				49
	1	Convert between gallons, quarts, pints, and cups	58-60	59-60	50-51
	2	Add and subtract gallons, quarts, pints, and cups	60-61	61	52
7	3	Practice	62		53
		Extra Practice, Unit 8, Exercise 2, pp. 143-146			
		Tests, Unit 8, 2A and 2B, pp. 85-90			
	Review 8		63-65	62-66	54
		Tests, Units 1-8, Cumulative A and B, pp. 91-104			
	Answers for unit 8 workbook exercises and review				55-56

Week		Objectives	Text book	Work book	Guide
Unit 9: Money					
	Chapter 1: Dollars and Cents				57
	1	Review: Money	66-67	67-68	58
		Extra Practice, Unit 9, Exercise 1, pp. 151-152			
	2	Make change, practice	68		59
		Tests, Unit 9, 1A and 1B, pp. 105-111			
	Chapter 2: Addition				60
8	1	Add Money	69-71	69-71	61
	2	Word problems	71-72	72-73	62
		Extra Practice, Unit 9, Exercise 2, pp. 153-158			
		Tests, Unit 9, 2A and 2B, pp. 113-118			
	Chapter 3: Subtraction				63
	1	Subtract Money	73-75	74-76	64-65
	2	Word problems	75-76	77-78	66
		Extra Practice, Unit 9, Exercise 3, pp. 159-162			
	3	Practice	77	79-80	67
		Tests, Unit 9, 3A and 3B, pp. 119-125			
	Chapter 4: Multiplication and Division				68
9	1	Multiply money	78-80	81-82	69
	2	Divide money	78, 80-81	83-84	70
	3	Practice	82		71
		Extra Practice, Unit 9, Exercise 4, pp. 163-164			
		Tests, Unit 9, 4A and 4B, pp. 127-132			
	Review 9		83-84	85-89	72
		Tests, Units 1-9, Cumulative A and B, pp. 133-144			
	Answers for unit 9 workbook exercises and review				73-75
Unit 10: Fractions					
	Chapter 1: Fraction of a Whole				76-77
10	1	Review: Fractions	85-87	90-95	78
		Extra Practice, Unit 10, Exercise 1A, pp. 175-176			
	2	Compare fractions	88-89	96-99	79-80
		Extra Practice, Unit 10, Exercise 1B, pp. 177-178			
	3	Practice	90		81
		Tests, Unit 10, Chapter 1, A and B, pp. 145-149			
	Chapter 2: Equivalent Fractions				82
	1	Understand equivalent fractions	91-92	100-101	83
	2	Use multiplication to find equivalent fractions	93	102-103	84

Week		Objectives	Text book	Work book	Guide
11	3	Use division to find equivalent fractions	94	104-105	85
	4	Find the simplest form of a fraction	94-95	106-107	86
		Extra Practice, Unit 10, Exercise 2, pp. 181-184			
	5	Compare fractions	95	108	87-88
	6	Practice	96		89
		Tests, Unit 10, 2A and 2B, pp. 151-157			
	Chapter 3: Adding Fractions				90
	1	Add like fractions	97-98	109-111	91
		Extra Practice, Unit 10, Exercise 3, pp. 185-186			
		Tests, Unit 10, 3A and 3B, pp. 159-165			
	Chapter 4: Subtracting Fractions				92
12	1	Subtract like fractions	99-100	112-114	93
		Extra Practice, Unit 10, Exercise 4, pp. 187-188			
	2	Practice	101		94
		Tests, Unit 10, 4A and 4B, pp. 167-174			
	Chapter 5: Fraction of a Set				95
	1	Find the fraction of a set	102-104	115	96
	2	Find the value of a fraction of a set	104	116	97
		Extra Practice, Unit 10, Exercise 5, pp. 189-190			
		Tests, Unit 10, 5A and 5B, pp. 175-182			
	Chapter 6: Fractions and Money				98
	1	Find the fraction of a dollar for a set of one kind of coin	105-107		99-100
13	2	Find the fraction of a dollar for a set of coins less than $1	107	117	101
		Extra Practice, Unit 10, Exercise 6, pp. 191-192			
	3	Practice	108		102
		Tests, Unit 10, 6A and 6B, pp. 183-188			
	Review 10		109-111	118-122	103
		Tests, Units 1-10, Cumulative A and B, pp. 189-199			
	Answers for unit 10 workbook exercises and review				104-109
Unit 11: Time					
	Chapter 1: Hours and Minutes				110
	1	Tell time to the 1 minute interval	112-113	123-124	111-112
14	2	Find elapsed time using clock faces	114-115	125-126	113
	3	Convert between hours and minutes	116	127-128	114

Week		Objectives	Text book	Work book	Guide
	4	Solve word problems involving time	117	129-130	115
		Extra Practice, Unit 11, Exercise 1A, pp. 197-200			
	5	Add and subtract time in hours and minutes	118-119	131-132	116-117
		Extra Practice, Unit 11, Exercise 1B, pp. 201-204			
	6	Practice	120		118
		Tests, Unit 11, 1A and 1B, pp. 201-208			
	Chapter 2: Other Units of Time				119
15	1	Measure time in seconds	121-123	133-136	120
	2	Convert between years and months	123	137-138	121
	3	Convert between weeks and days	123	139-140	122
		Extra Practice, Unit 11, Exercise 2, pp. 205-206			
	4	Practice	124		123
		Tests, Unit 11, 2A and 2B, pp. 209-214			
	Review 11		125-126	141-145	124
		Tests, Units 1-11, Cumulative, A and B, pp. 215-226			
	Answers for unit 11 workbook exercises and review				125-127
Unit 12: Geometry					
	Chapter 1: Angles				128
16	1	Identify angles and polygons	127-129	146-147	129-130
		Extra Practice, Unit 12, Exercise 1, pp. 211-212			
		Tests, Unit 12, 1A and 1B, pp. 227-234			
	Chapter 2: Right Angles				131
	1	Identify right angles	130-131	148-150	132
		Extra Practice, Unit 12, Exercise 2, pp. 213-214			
		Tests, Unit 12, 1A and 1B, pp. 235-242			
	Chapter 3: Quadrilaterals and Triangles				133-134
	1	Identify quadrilaterals and triangles	132-134	151-152	135-136
		Extra Practice, Unit 12, Exercise 3, pp. 215-216			
		Tests, Unit 12, 1A and 1B, pp. 243-250			
	Chapter 4: Solid Figures				137
	1	Identify solids	135-136	153-154	138-139
		Extra Practice, Unit 12, Exercise 4, pp. 217-218			
		Tests, Unit 12, 4A and 4B, pp. 251-258			
	Review 12		137-138	155-158	140
		Tests, Units 1-12, Cumulative A and B, pp. 259-272			
	Answers for unit 12 workbook exercises and review				141-143

Materials

Whiteboard and Dry-Erase Markers

Multilink cubes
These are cubes that can be linked together on all 6 sides.

Number cubes
Cubes that you can label and throw, like dice. You need two. You can use regular dice and label them with masking tape, or buy cubes and labels.

Measuring tools
Meter stick.
Ruler marked in centimeters and inches.
Measuring tape marked in centimeters and inches.
A simple pan or bucket balance.
A kitchen scale showing kilograms and grams as well as pounds and ounces.
Kilogram, gram, pound, and ounce weights.
Liter/Quart measuring cup marked in milliliters as well as cups.
Cup or Pint measuring cups also marked in milliliters.
Optional: beakers of various capacitates, e.g. liter, 200 ml, 100 ml.
Optional: medicine spoon marked in milliliters.

Coins and Bills
Play or real money. Bills up to $100 and coins. If you don't have many bills, write $1, $5, $10, $20, $50, and $100 on paper about the size of bills.

Store cards
Cards with pictures of items with costs up to $100. You can use pictures from magazines, newspaper ads, coupons, etc.. Glue to index cards, and write a cost, using decimal notation, e.g. $4.60. Have an equal number of items for prices less than $1, less than $5, less than $20, less than $50, and less than $100.

Fraction bars and circles
You can copy the ones in the appendix or use commercially available fraction bars and circles.

Counters
Round counters are easy to use and pick up, but any type of counter will work.

Clock
Two face-clocks with geared hands. A large demonstration clock is nice to teach with but not required.

Index cards

Solid shapes
Cubes, prisms, pyramids, cylinders, cones, and spheres. Preferably of similar sizes that the student can hold, examine, and compare.

Centimeter cubes

Supplements

The textbook and workbook provide the essence of the math curriculum. Some students profit by additional practice or more review. Other students profit by more challenging problems. There are several supplementary workbooks available at www.singaporemath.com.

If you feel it is important that your student have a lot of drill in math facts, there are many websites that generate worksheets according to your specifications, or provide on-line fact practice. Web sites come and go, but doing a search using the terms "math fact practice" will turn up many sites. Playing simple games is another way to practice math facts.

Unit 6 – Length

Chapter 1 – Meters and Centimeters

Objectives

- Review meters and centimeters as units of length.
- Estimate and measure lengths in meters and centimeters.
- Convert a measurement in meters and centimeters to centimeters
- Convert a measurement in centimeters to meters and centimeters.
- Add and subtract lengths in meters and centimeters.

Vocabulary

- Centimeter
- Meter
- About
- Convert
- Unit conversion factor
- Precise
- Compound units
- Metric system

Notes

In *Primary Mathematics* 2A, students learned to measure length in meters *or* centimeters. This is reviewed briefly in this chapter. Then students will measure length in meters *and* centimeters. They will also convert between measurements. Students have already converted money in compound units. Converting between meters and centimeters is very similar, since 1 meter = 100 centimeters.

Students will be learning to measure in two distinct measurement systems, the metric system and the U.S. customary system. When they measure in compound units, they will be measuring length in one or the other system and should not mix them; for example, they should not measure a length as 1 meter 4 inches.

The metric system has become the global language of measurement and today about 95% of the world's population uses the metric system. The use of the metric system is legal (but not mandatory) in the United States. The metric system is used in science.

The metric system of measurement is based on powers of 10. The prefixes tell what multiple of the basic unit is being considered. For length, the SI unit (International Standards unit) is a meter.

*Kilo*meter	1000 meters
*Hecto*meter	100 meters
*Deca*meter	10 meters
Meter	
*Deci*meter	0.1 meters
*Centi*meter	0.01 meters
*Milli*meter	0.001 meters

Since objects do not always measure to exactly a certain number of centimeters, your student must measure to the *nearest* centimeter. This idea will be built upon later in chapters dealing with estimation. Use the term **about** when saying that a length is close to a certain number of meters or centimeters. You can also teach your student to measure to the nearest millimeter, but that is not required at this level.

Your student should have plenty of practical experience measuring lengths. Make sure he measures from the 0 mark on the ruler or measuring tape. In order to be able to estimate lengths, he should have some idea of the length of his arms, hands, or feet or width of his palms or fingers. For example, if the width of one of his fingers is about 1 cm, he can visualize the number of fingers-wide an object is.

In this and subsequent chapters, your student will learn to add and subtract in compound units. Although addition and subtraction of compound units can be performed by converting the bigger unit to the smaller unit, such as converting meters and centimeters to centimeters only, and then adding or subtracting with the formal algorithm, encourage your student to use mental strategies. Your student has already learned in previous levels to make the next hundred or subtract from hundreds. Subtracting centimeters from meters is the same as subtracting from hundreds. The mental math strategies used in this particular chapter are very similar to those that were learned when adding and subtracting dollars and cents in *Primary Mathematics* 2B.

Material

- Meter stick
- Ruler
- Measuring tape
- 2 number cubes
- Mental Math 1 (Appendix)

(1) Measure in meters and centimeters

Discussion

Concept page 8

Point out the abbreviations for meter (m) and centimeter (cm). Tell your student that the meter and centimeter are part of the metric system of measurement that is used in most of the world.

Activity

If your student has not used earlier levels of *Primary Mathematics* and has not measured lengths in meters and centimeters, spend some time familiarizing her with the length of a centimeter and a meter. Below are some possible activities you can ask your student to do.

⇒ Measure some objects to the nearest centimeter.

⇒ Estimate the lengths of various objects or lines, and then measure.

⇒ Estimate and then find out if the door is about 1 m, 2 m, or 3 m tall. Is it about 1 m, 2 m, or 3 m wide.

⇒ Measure the length and width of your thumb to the nearest centimeter.

⇒ Find the distance straight across from the tip of your thumb to the tip of your index finger when your thumb is spread out from the rest of your hand.

⇒ Find the distance from your elbow to the tip of your fingers.

⇒ Find something that is about 5 cm long. Find something about 10 cm long.

⇒ See how close to 8 cm can you cut a piece of string without measuring first.

⇒ Use a ruler to draw a line 3 cm long.

⇒ Without a ruler, draw a line 5 cm long. Then measure to see how close you came.

Discussion

Tasks 1-2, p. 9

You can have your student first estimate the length or width of the items when appropriate in tasks 1 and 2 and then measure them to verify their estimate.

1. (a) meter stick
   (b) measuring tape
   (c) ruler
   (d) ruler/meter stick
2. (a) meters
   (b) centimeters
   (c) centimeters
   (d) meters

Activity

Have your student look at the meter stick and see that there are 100 centimeters in a meter.

Ask your student for the number of centimeters in 2 m, 3 m, or 5 meters. Write the multiplication equations. Tell your student that 100 is the *conversion factor* for going from 1 meter to centimeter. To *convert* a measurement in meters to centimeters, we just multiply by the *conversion factor* for meters to centimeters.

1 m = 100 cm
2 m = 100 cm x 2 = 200 cm
3 m = 100 cm x 3 = 300 cm
5 m = 100 cm x 5 = 500 cm

Have your student measure something that is about a meter and a half long. Tell him that rather than say the objects is about 1 meter or about 2 meters, we can be more *precise* in our measurement by measuring to the nearest centimeter. We can then give the measurement as 1 meter along with the number of centimeters, such as 1 m 30 cm.

Ask your student to convert 1 m 30 cm to centimeters only. Since 1 m is 100 cm, 1 m 30 cm is 100 and 30, or 130 cm. Have him convert some other measurements, including ones where the centimeters are less than 10.

1 m 30 cm = 100 cm + 30 cm = 130 cm
2 m 64 cm = 200 cm + 64 cm = 264 cm
6 m 5 cm = 600 cm + 5 cm = 605 cm
50 m 25 cm = 5000 cm + 25 cm = 5025 cm

Write 100 cm = 1m. Then ask her how many meters is 200, 300, 500 cm.

Then, write down 250 cm and ask your student to tell you how many meters and centimeters this is. To find the answer, we can split out hundreds and convert them into meters.

Have your student convert some other amounts to meters and centimeters, including some where the centimeters will be less than 10.

100 cm = 1 m   200 cm = 2 m
300 cm = 3 m
500 cm = 5 m

250 cm
/ \
200 cm   50 cm
= 2 m
\ /
2 m 50 cm

360 cm = 3 m 60 cm
915 cm = 9 m 15 cm
704 cm = 7 m 4 cm

Tell your student than when we give a measurement with more than one unit, such as meters and centimeters, we will say that the measurement is in *compound units*. When we use compound units, we will always convert completely. For example 7 m 345 cm should not be given as an answer, even though it is a correct measurement, because 300 of the centimeters are 3 meters. The answer should be given as 10 m 45 cm.

Practice

Tasks 3-10, pp. 9-10

Workbook

Exercise 1, p. 7 (answers p. 24)

Reinforcement

Extra Practice, Unit 1, Exercise 1A, pp. 5-6

3. (a) 25 cm
   (b) 125 cm
4. (a) 200 cm
   (b) 3 cm
5. Answers will vary.
6. 145 cm
7. (a) 190 cm   (b) 155 cm   (c) 286 cm
   (d) 289 cm   (e) 308 cm   (f) 406 cm
8. 3 m 95 cm
9. (a) 1 m 80 cm   (b) 1 m 95 cm   (c) 2 m 62 cm
   (d) 2 m 99 cm   (e) 3 m 4 cm   (f) 4 m 9 cm
10. 1 m 89 cm, 1 m 96 cm, 2 m 8 cm

(2) Subtract from meters

Discussion

Tasks 11-12, p. 11

11: Since 1 m = 100 cm, we can subtract 35 cm from 1 m by subtracting 35 from 100. Use these tasks to remind your student of mental math strategies for subtracting from 100, as shown with task 11. We can count up by ones to the next ten and then by tens to 100, or by tens to a number between 90 and 100 and then by ones to 100. Or, we can think of 100 as 9 tens and 10 ones, and subtract tens from tens and ones from ones.

12: 2 m 35 cm is between 2 m and 3 m, so the difference can be found by subtracting 35 cm from 100 cm

11. 65 cm
12. 65 cm

Practice

Task 13, p. 11

13. (a) 60 cm (b) 15 cm
(c) 57 cm (d) 33 cm
(e) 68 cm (f) 9 cm
(g) 75 cm
(h) 54 cm
(i) 96 cm
(j) 27 cm

Workbook

Exercise 2, p. 8 (answers p. 24)

Reinforcement

Mental Math 1

Label two number cubes, one with 0-5, and the other with 4-9. Roll the two cubes. Allow your student to pick one to be the tens and the other to be ones. For example, if a 6 and a 2 is rolled the number can be 62 or 26. Your student must give the number that makes 100 with this number.

(3) Add and subtract meters and centimeters

Activity

Write the following problems and discuss strategies for solving them. All answers should be in meters and centimeters.

⇒ 83 cm + 29 cm

We can add using mental math strategies, or rewrite the problem vertically and use the addition algorithm. Then convert to meters and centimeters.

83 cm + 29 cm
83 + 29 = 112
83 cm + 29 cm = 1 m 12 cm

⇒ 85 cm + 30 cm

We can "make 100" or "make 1 m" by taking 15 from the 30, which leaves 15 cm.

85 cm + 30 cm = 1 m 15 cm
(30 → 15 15)

⇒ 4 m 85 cm + 30 cm

We can "make the next meter" by taking 15 from the 30, or add the centimeters and convert, giving one more meter.

4 m 85 cm + 30 cm = 5 m 15 cm
(30 → 15 15)

⇒ 32 m 75 cm + 2 m 28 cm

[1] Add the meters first.

[2] Then add the centimeters, either by mentally adding 75 and 28 to get 103 cm, converting to 1 m 3 cm, and adding that to the 34 m, or by making a meter with the 75 cm.

Or, convert both measurements to centimeters, use the addition algorithm, and convert back to meters and centimeters.

32 m 75 cm + 2 m 28 cm

[1] 32 m 75 cm + 2 m = 34 m 75 cm

[2] 34 m 75 cm + 28 cm = 34 m + 103 cm
= 35 m 3 cm

Or
34 m 75 cm + 28 cm = 35 m 3 cm
(28 → 25 3)

Or
```
  1 1
  3 2 7 5
+   2 2 8
  3 5 0 3  →  35 m 3 cm
```

⇒ 4 m 30 cm – 48 cm

Subtract 48 cm from one of the meters and add back in the 30 cm. Or use the subtraction algorithm.

4 m 30 cm – 48 cm = 3 m 82 cm
(4 m → 3 m 100 cm)
100 – 48 = 52; 52 + 30 = 82

Or
```
  4 3 0
–   4 8
  3 8 2  →  3 m 82 cm
```

⇒ 44 m 2 cm – 16 m 64 cm

[1] Subtract the meters first.

[2] Then, subtract the centimeters, either by subtracting from a meter (making 100 with 62) and adding back in the 2 cm, or by subtracting 64 from 102.

Or, convert both measurements to centimeters, subtract vertically, and convert back to meters and centimeters.

44 m 2 cm – 16 m 64 cm

[1] 44 m 2 cm – 16 m = 28 m 2 cm

[2] 28 m 2 cm – 64 cm = 27 m 38 cm

27 m 100 cm

100 – 64 = 36; 36 + 2 = 38

Or

$$\begin{array}{r} 102 \\ -\ 64 \\ \hline 38 \end{array}$$

Or

$$\begin{array}{r} \overset{3}{\cancel{4}}\,\overset{^{1}3}{\cancel{4}}\,\overset{9}{\cancel{0}}\,^{1}2 \\ -\ 1\,6\,6\,4 \\ \hline 2\,7\,3\,8 \end{array}$$

2738 → 27 m 38 cm

Discussion

Task 14, p. 12

You can discuss different strategies for adding and subtracting the 85 cm.

14. (a) 5 m 25 cm
(b) 1 m 55 cm

Practice

Task 15, p. 12

15. (a) 9 m 32 cm (b) 3 m 79 cm
(c) 8 m 30 cm (d) 6 m 46 cm
(e) 3 m 20 cm (f) 9 m 1 cm
(g) 6 m 15 cm (h) 5 m 24 cm
(i) 4 m 55 cm (j) 3 m 95 cm
(k) 3 m 55 cm (l) 2 m 30 cm

Workbook

Exercise 3, pp. 9-10 (answers p. 24)

Reinforcement

Extra Practice, Unit 6, Exercise 1, pp. 113-114

Test

Tests, Unit 6, 1A and 1B, pp. 1-6

Chapter 2 – Kilometers

Objectives

- Understand the kilometer as a unit of measurement.
- Convert a measurement in kilometers and meters to meters.
- Convert a measurement in meters to kilometers and meters.
- Add and subtract lengths in kilometers and meters.

Vocabulary

- Kilometers
- Kilo-
- Centi-

Notes

Kilometers are used to measure longer distances in the metric system.

1 kilometer = 1000 meters

In the U.S., miles are used to measure longer distances, and your student will probably be more familiar with miles. A mile is a little over one and a half kilometers (1 mile = 1.6093 km) or a kilometer is a little less than two thirds of a mile (1 km = 0.6214 miles). So 6 miles is about 10 km, and 60 miles is about 100 km. A car traveling at 100 km per hour is traveling at about 60 miles per hour.

If your student is familiar with a distance about a mile away, you can use that to give her a feel for kilometers. For example, if the post office is a little over a mile away, it is about 2 kilometers away. Something that takes about an hour to reach in the car is about 100 kilometers away. If your student is familiar with distances on a map for places she has been to frequently, you may want to determine those distances in kilometers and use the map to help her become more familiar with kilometers.

Later, students will learn that a liter is 1000 milliliters, and a kilogram is 1000 g. Since 1 kilometer is 1000 meters, the more common conversion factor will be 1000. However, 1 meter is 100 centimeters. To help your student remember whether to use 1000 or 100, you can teach him that kilo- means a thousand, and centi- means 100. A *kilo*meter is 1000 meters, a meter is 100 *centi*meters.

Material

- Meter stick
- Number cards, 4 sets 0-9
- Mental Math 2-4 (Appendix)

(1) Measure in kilometers and meters

Discussion

Concept page p. 13

Tell your student that longer distances are measured in kilometers. You can show her a meter stick and have her imagine 10 of them placed end to end. If possible, you can measure a distance of 10 meters from a wall and down a hallway. Then, ask her to imagine 100 of those distances, or 1000 meter sticks end to end. That is 1 kilometer. Discuss other distances that are about a kilometer, such as from the house to a specific intersection or another house.

Point out that the abbreviation for kilometers is km, and that 1 km = 1000 m. Tell your student that the prefix *kilo-* means 1000. Remind him that 1 m = 100 centimeters. The prefix *centi-* means 100, as in centipede or cents. Remembering that *centi-* means 100 and *kilo-* means 1000 will make it easier to remember that 1 kilometer is 1000 meters, but 1 meter is 100 centimeters.

Activity

Write down 1 km = 1000 m and then have your student find the number of meters in 2 km, 3 km, and 5 km. We simply multiply the number of kilometers by 1000 to convert to meters. Then have her find the number of meters in 1 km 500 m, 1 km 50 m, and 1 km 5 m.

1 km = 1000 m
2 km = 2 x 1000 m = 2000 m
3 km = 3 x 1000 m = 3000 m
5 km = 5 x 1000 m = 5000 m

1 km 500 m = 1500 m
1000 m   500 m

1 km 50 m = 1050 m
1000 m   50 m

1 km 5 m = 1005 m
1000 m   5 m

Write down 1000 m = 1 km and have your student find the number of kilometers in 4000 m, 6000 m, and 10,000 m. Each thousand is a kilometer. Then have him find the number of kilometers and meters in 2500 m, 2050 m, and 2005 m.

1000 m = 1 km
4000 m = 4 km
6000 m = 6 km
10,000 m = 10 km

2500 m = 2 km 500 m
2 km   500 m

2050 m = 2 km 50 m
2 km   50 m

2005 m = 2 km 5 m
2 km   5 m

Discussion

Tasks 1-4, pp. 14-15

1. (a) 1 km 10 m
   (b) 1 km 750 m
2. (a) 42 km; 23 km
   (b) 41 km
3. 6100 m
4. 400 m x 3 = 1200 m = 1 km 200 m

Practice

Tasks 5-6, p. 15

5. (a) 1600 m (b) 2550 m (c) 2605 m
   (d) 3085 m (e) 3020 m (f) 4005 m
6. (a) 1 km 830 m (b) 2 km 304 m (c) 2 km 780 m
   (d) 3 km 96 m (e) 3 km 40 m (f) 4 km 9 m

Workbook

Exercise 4, pp. 11-14 (answers p. 24)

For problems 8(b-c), you ask your student to give the answers in compound units, since the problems do not specify compound units.

(2) Subtract from kilometers

Discussion

Tasks 7-10, p. 16

Since 1 km is 1000 m, we can subtract meters from a kilometer by subtracting the meters from 1000 m. These tasks show a mental math strategy for subtracting from 1000. We can think of 1000 as 9 hundreds, 9 tens, and 10 ones. So to subtract from 1000, we subtract the digit in the hundreds place from 9, the digit in the tens place from 9, and the digit in the ones place from 10. If the digit in the ones place is 0, then we think of 1000 as 9 hundreds and 10 tens. Go through these tasks with your student, making sure he understands the strategy. If you want him to have more practice with subtracting from 1000 before continuing, you can have him do Mental Math 2.

7. 650 m
   650 m
   650 m
8. 995 m
9. 645 m
10. 375 m

Tasks 11-12, p. 17

These tasks show that when we are subtracting from a kilometer and meters, or adding meters to meters, we can use the standard algorithm to add or subtract when it is not easy to use mental math.

11. 235 m
12. 2335 m

Activity

Write down the following problems and discuss strategies for solving them using mental math strategies.

⇒ 1 km 300 m – 600 m

In this case, we can use mental math strategies, because it is easy to subtract 600 m from 1 km and then add back in the 300 m.

⇒ 1 km 830 m + 400 m

This could also be done mentally by "making 1000."

⇒ 1 km 99 m + 345 m

Here, we can use the strategy learned for adding a number close to 100. We can add 99 to 345 by adding 100 and subtracting 1.

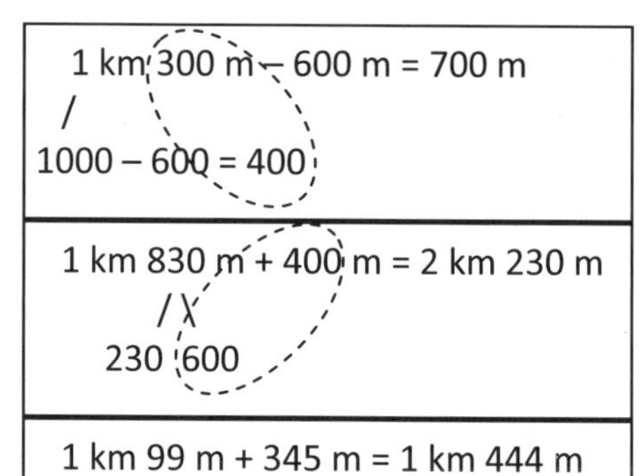

Practice

Task 13, p. 17

13. (a) 760 m (b) 752 m
    (c) 955 m (d) 993 m
    (e) 676 m (f) 99 m
    (g) 875 m
    (h) 955 m
    (i) 1 km 943 m
    (j) 5 km 2 m

Workbook

Exercise 5, pp. 15-16 (answers p. 25)

Reinforcement

Mental Math 2-4

(3) Add and subtract kilometers and meters

Activity

Write the following problems and discuss strategies for solving them.

⇒ 600 m + 580 m

Add the meters and then convert to kilometers and meters. Or, make 1 km using 400 of the 580 m.

600 m + 580 m = 1180 m = 1 km 180 m
Or
600 m + 580 m = 1 km 180 m
/\\
400 180

⇒ 2 km 965 m + 6 km 85 m

[1] Add the kilometers.

[2] Add the meters and then convert to kilometers and meters, if there are more than 1000 m.

Or

Use mental math to make 1 km.

Or

Convert both measurements to meters, use the addition algorithm, and convert back to kilometers and meters.

2 km 965 m + 6 km 85 m

[1] 2 km 965 m + 6 km = 8 km 965 m

[2] 8 km 965 m + 85 m
= 8 km + 1050 m
= 9 km 50 m

$$\begin{array}{r} 965 \\ +\ \ 85 \\ \hline 1050 \end{array}$$

or

8 km 965 m + 85 m = 9 km 50 m
/\\
35 50

or

$$\begin{array}{r} 2965 \\ +\ 6085 \\ \hline 9050 \end{array} \rightarrow 9 \text{ km } 50 \text{ m}$$

⇒ 5 km 130 m – 25 m

We simply subtract meters from meters.

5 km 130 m – 25 m = 5 km 105 m

⇒ 5 km 130 m – 300 m

We can rename 1 km as 1000 m and subtract the meters from 1130 m.

Or

We can subtract 300 m from 1 km, and add back in the 130 m.

5 km 130 m – 300 m

4 km 1130 m – 300 m = 4 km 830 m

or

5 km 130 m – 300 m = 4 km 830 m
/\\
4 km 1000 m
1000 m – 300 m = 700 m
700 m + 130 m = 830 m

⇒ 9 km 5 m – 4 km 450 m

[1] Subtract the kilometers.

[2] Subtract the meters and then convert to kilometers and meters, if there are more than 1000 m.

Or

Use mental math and subtract from 1 km.

Or

Convert both measurements to meters, add, and convert back to kilometers and meters.

9 km 5 m – 4 km 450 m

[1] 9 km 5 m – 4 km = 5 km 5 m

[2] 5 km 5 m – 450 m
= 4 km 1005 m – 450 m
= 4 km 555 m

```
  1 0 0 5
-   4 5 0
    5 5 5
```

5 km 5 m – 450 m = 4 km 555 m
/ \
4 km 1000 m
1000 m – 450 m = 550 m
550 m + 5 m = 555 m

or

```
  9 0 0 5
- 4 4 5 0
  4 5 5 5   →   4 km 555 m
```

Discussion

Tasks 14-15, p. 18

You can discuss different strategies for adding 850 m and subtracting 920 m.

14. 4 km 300 m
4 km 300 m

15. 120 m
120 m

Practice

Task 16, p. 18

16. (a) 9 km 325 m (b) 3 km 890 m
(c) 8 km 100 m (d) 6 m 245 m
(e) 6 km 150 m (f) 5 km 800 m
(g) 4 km 600 m (h) 3 km 950 m
(i) 3 km 550 m (j) 2 km 255 m

Workbook

Exercise 6, pp. 17-18 (answers p. 25)

Reinforcement

Extra Practice, Unit 6, Exercise 2, pp. 115-116

(4) Practice

Practice

Practice A, p. 19

1. (a) 400 cm	(b) 140 cm	(c) 225 cm
2. (a) 1 m 20 cm	(b) 2 m 25 cm	(c) 3 m 9 cm
3. (a) 35 cm (b) 25 cm (c) 8 cm		
4. (a) 5 m 75 cm (c) 5 m 80 cm	(b) 3 m 69 cm (d) 5 m 14 cm	
5. 1 m 60 cm – 16 cm = 1m 44 cm Ryan is **1 m 44 cm** tall.		
6. 3000 m	(b) 1450 m	(c) 2506 m
7. (a) 1 km 680 m	(b) 1 km 85 m	(c) 2 km 204 m
8. (a) 200 m (b) 955 m (c) 60 m		
9. (a) 5 km 650 m (c) 1 km 950 m	(b) 4 km 100 m (d) 2 km 675 m	

Reinforcement

Get your student to make up a map, perhaps of an island or a treasure map, and mark reasonable distances. Then ask her to make up word problems based on the distances on the map.

Test

Tests, Unit 6, 2A and 2B, pp. 7-12

Game

Material: 4 sets of number cards 0-9.

Procedure: Shuffle and place face down in the middle. Players turn over 3 cards, form a 3-digit number with it, subtract that number from 1000, and write the difference down. Players then turns over another 3 cards, and repeat the process, writing down another difference. They then add the two numbers together. The player with the highest number wins the round.

Chapter 3 – Yards, Feet and Inches

Objectives

- Measure and estimate lengths in yards, feet, and inches.
- Compare yard to meter and inch to centimeter.
- Convert a measurement between yards and feet, and between inches and feet.
- Add and subtract yards and feet in compound units.
- Add and subtract feet and inches in compound units.

Vocabulary

- Yards
- Foot
- Feet
- Inches
- Customary system of measurement

Notes

In *Primary Mathematics* 2A students learned to measure in yards, feet, and inches. In this chapter they will measure in compound units, learn to convert between yards, feet, and inches, and add and subtract in compound units.

Students will not be converting between measurement systems in *Primary Mathematics*, but it is useful for them to have an approximate idea of how they compare.

1 inch = 2.54 cm ≈ two and a half centimeters
1 foot = 30.48 cm ≈ 30 cm
1 yard = 0.9144 m ≈ 1 meter
1 meter = 1.0936 yards ≈ 1 yard (≈ 40 in.)

Although students can convert measurements to the smaller unit, then add and subtract normally, and then convert back to compound units, conversions in this system are cumbersome. Students will learn some mental math strategies similar to ones used for adding and subtracting compound units in the metric system, except that 1 yard is renamed as 3 feet, and 1 foot is renamed as 12 inches. Renaming in a base other than 10 is good preparation for other instances where the base is not ten; for example, when adding or subtracting hours and minutes or when adding or subtracting whole numbers and fractions. We can "make a 3" or "subtract from a 3" in the case of yards and feet, and "make a 12" or "subtract from a 12" in the case of feet and inches.

If your student has difficulties applying mental math strategies to U.S. measurements in compound units, you can have him convert to the smaller unit, add or subtract using the standard algorithm or mental math strategies for 2-digit and 3-digit numbers he has already learned, and then convert back to a compound unit.

Material

- Yard stick
- Ruler
- Meter stick
- Measuring tape
- Appendix pp. a7-a8
- Mental Math 5-6 (Appendix)

(1) Convert between yards and feet

Discussion

Concept page 20

Use the activities on this page to review yards, feet, and inches.

Point out the abbreviation for yard, foot, and inch at the bottom of the page. The abbreviation for more than one foot, or feet, is the same as for foot. For the abbreviation of inches, we put a period at the end to tell it apart from the word *in*.

Tell your student that yards, feet, and inches are part of the customary system of measurement used in the United States, and only a few places elsewhere in the world. Sometimes it is called the Imperial system or English system.

Activity

You may want to have your student also do some of the following activities.

⇒ Have your student measure some objects to the nearest inch, half-inch, and quarter-inch. Show your student the half-inch and quarter-inch marks on the ruler.

⇒ Tell your student that, as with meters and centimeters, we can measure lengths in yards and feet, or feet and inches, or yards and inches, or even yards, feet, and inches. Have him use a yard stick and rulers by marking the end of a yard and moving the yard stick along, and then using a ruler when the rest of the distance is less than a yard. Similarly, he can measure distances in feet and inches. (Don't have him convert measurements at this point.)

⇒ Have your student measure his height in feet and inches.

⇒ Ask your student to find a body part, such as the length from a knuckle to the tip of the finger, that is about an inch long.

⇒ Ask your student to draw a line 3 inches long without a ruler and then measure to see how close he came.

⇒ Ask your student to use a ruler to draw a line 5 inches long, estimate its length in centimeters, and then check.

Have your student do the following activities to relate U.S. customary measurement units to metric units. Use a yard stick, meter stick, rulers, or a measuring tape showing both inches and centimeters to answer the following:

⇒ Which is longer, an inch or a centimeter? (inch)

⇒ About how much longer? (a little more than twice as long)

⇒ About how many inches are there in 10 cm? (About 4)

⇒ In 20 cm? (About 8)

⇒ In 30 cm? About 12)

⇒ About how many centimeters are there in a foot? (About 30)

⇒ Which is longer, a meter or a yard? (meter)

⇒ How much longer? (about 9 cm or about 3 inches longer)

⇒ How many feet are there in a yard? (1 yard = 3 feet)

Discussion

Tasks 1-5, pp. 20-21

1: Since 1 yard is 3 feet, 1 yd 2 ft is 2 ft longer than 3 ft.

2: We convert yards to feet by multiplying the number of yards by the conversion factor 3.

3-4: We divide by 3 to convert from feet to yards. You may want to tell your student that we can write the conversion factor as ft/yd, which we read as "feet in a yard" or "feet per yard." So 3 ft/yd means there are 3 ft in 1 yard.

5: If the number of feet is not divisible by 3, then we can split out the part that is divisible by 3, convert it to yards, and the remainder is the feet. This is the same process as dividing with remainders. The quotient is the number of yards, and the remainder is the number of feet. (Note: The first printing of the textbook has an error, the second line in the second blue box should be 25 **ft** = 8 yd 1 ft.)

1. (a) 2 ft
   (b) 3 ft + 2 ft = **5** ft
2. 11 yd 2 ft
   11 x 3 = 33
   11 yd 2 ft = 33 ft + 2 ft = **35 ft**
3. (a) 8 yd x 3 ft/yd = **24** ft
   (b) 18 ft ÷ 3 ft/yd = **6** yd
4. 15 ft ÷ 3 ft/yd = **5** yd
5. (a) 20 ÷ 3 = 6 R 2
   20 ft ÷ 3 ft/yd = **6** yd **2** ft
   (b) 25 ft = **8** yd **1** ft
   (c) 422 ft = **140** yd **2** ft

```
      1 4 0
  3 ) 4 2 2
      3
      1 2
      1 2
        0 2
```

Activity

To help your student remember whether to multiply or divide when converting measurements, discuss the following concept. Since 1 yard is the same as 3 feet, when we go from yards to feet, the number will be greater. For example, 2 yards is the same length as 6 feet, but 2 is a smaller number than 6. This is because feet are shorter units than yards. So when we convert from yards to feet, we are going to need more feet for the same measurement. Do we get more if we multiply, or if we divide? We get more if we multiply.

When we go from feet to yards, we are putting three feet together to make a yard. We need a smaller number for the same measurement. So we divide.

You can use a bar-model type of drawing to illustrate this.

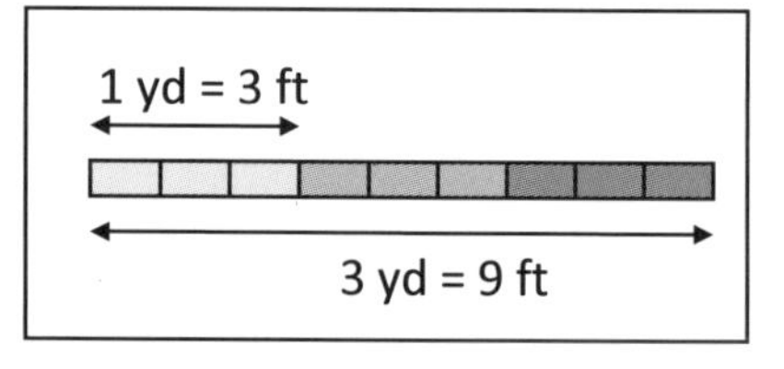

Practice

Appendix p. a7

1. (a) >   (b) =
   (c) >   (d) <
   (e) >   (f) =
2. (a) 320 + 342 = 662
   662 ÷ 3 = 220 R 2
   220 yd 2 ft
   (b) 66 yd x 3 ft/yd = 198 ft
   198 ft + 478 ft = **676 ft**
   (c) 3 yd x 3 ft/yd = **9 ft**
   (d) 320 + 478 = 798
   798 ÷ 3 = 266
   266 yd 0 ft

(2) Convert between feet and inches

Activity

Have your student look at a ruler and tell you the number of inches in a foot. Write 1 ft = 12 in.

Tell your student that there are 12 inches in a foot. We can write this as 12 in./ft. Ask your student for the number of inches in 2 ft, 3 ft, and so on through 10 ft. Write the multiplication equation and the answer.

If your student has not yet memorized multiplication facts for 12, have him practice counting by 12's. He can count by 12's by adding ten and two ones each time. He can use mental math strategies for multiplying by 12: 12 x 8 = 80 + 16.

1 ft = 12 in.

1 ft = 1 x 12 in./ft = 12 in.
2 ft = 2 x 12 in./ft = 24 in.
3 ft = 3 x 12 in./ft = 36 in.
4 ft = 4 x 12 in./ft = 48 in.
5 ft = 5 x 12 in./ft = 60 in.
6 ft = 6 x 12 in./ft = 72 in.
7 ft = 7 x 12 in./ft = 84 in.
8 ft = 8 x 12 in./ft = 96 in.
9 ft = 9 x 12 in./ft = 108 in.
10 ft = 10 x 12 in./ft = 120 in.

Discussion

Tasks 6-8, pp. 21-22

6: Since 1 ft is 12 in., we can find 1 ft 5 in. by adding 5 to 12.

7: To convert a measurement in feet and inches to all inches, we multiply the number of feet by 12, and add on the inches. Point out that in going from feet to inches, we are essentially cutting up each foot into 12 smaller parts. So we are going from a smaller number of feet to a larger number of inches. Each foot is now multiplied by 12.

6. (a) 5 in.
   (b) 12 in. + 5 in. = **17** in.

7. 7 ft x 12 in./ft = **84** in.
   7 ft. 3 in. = 84 in. + 3 in. = **87** in.

8. Answers will vary.

Practice

Task 9, p. 22

9. (a) 6 yd x 3 ft/yd = **18** ft
   (b) 32 yd x 3 ft/yd = **96** ft
   (c) 9 yd 2 ft = 27 ft + 2 ft = **29** ft
   (d) 12 yd 1 ft = 36 ft + 1 ft = **37** ft
   (e) 30 ft ÷ 3 ft/yd = **10** yd **0** ft
   (f) 24 ft ÷ 3 ft/yd = **8** yd **0** ft
   (g) 123 ft ÷ 3 ft/yd = **41** yd **0** ft
   (h) 100 ft ÷ 3 ft/yd = **33** yd **1** ft
   (i) 1 ft 5 in. = 12 in. + 5 in. = **17** in.
   (j) 4 ft x 12 in./ft = **48** in.
   (k) 4 ft 9 in. = 48 in. + 9 in. = **57** in.
   (l) 6 ft 4 in. = 72 in. + 4 in. = **76** in.

Workbook

Exercise 7, pp. 19-20
(answers p. 25)

Reinforcement

Mental Math 5

Enrichment

Write down 24 in. and ask your student how many feet it is. Your student will probably be able to tell you it is 2 ft. Then write down 27 in. and ask her how many feet it is. It is 2 ft with 3 in. left over. Tell her that we can convert from inches to feet or to feet and inches by finding how many groups of 12 there are, and then finding the left-overs. She has not yet learned to divide by a 2-digit number, but she can convert up to 120 in. into feet and inches by using multiplication facts for 12, or counting by 12's to get the number of feet, or subtracting 12 repeatedly. Have her do the following problems.

24 in. = 2 ft.
27 in. = 2ft. 3 in.

30 in. = _____ ft _______ in. (2 ft 6 in.)
36 in. = _____ ft _______ in. (3 ft 0 in.)
41 in. = _____ ft _______ in. (3 ft 5 in.)

130 in. = _____ ft _____ in. (10 ft 10 in.)
Which is greater, 3 ft or 40 in.? (40 in.)
Which is greater, 2 yd 2 ft or 90 in.? (2 yd 2 ft)

(3) Add and subtract yards and feet in compound units

Discussion

Tasks 10-12, pp. 22-23

10: When we add feet, we can "make 3" to create a yard. Since compound measures generally already have all feet converted to yards that can be, there will only be 1 or 2 feet. So the only instances where "make 3" will occur is when both measurements have 2 ft or one is 2 ft and the other is 1 ft.

11: To add yards and feet, we can add the yards together, and then the feet, converting 3 ft into 1 yard. Your student can probably do this mentally without writing an intermediate step.

12: To subtract feet from yards, we know we have to subtract it from one of the yards. There will be one less yard, and the difference between the number of feet and 3 feet. Your student may see the similarity to "subtracting from a 10." Instead, we are "subtracting from a 3."

10. 1 yd 1 ft

11. 6 yd 1 ft

12. (a) 1 ft
(b) 11 yd 1 ft

3 yd 2 ft + 2 yd 2 ft = 6 yd 1 ft
1 1

Activity

Write the following expressions and guide your student in adding or subtracting in compound units.

⇒ 3 yd 2 ft + 2 yd 1 ft

3 yd 2 ft + 2 yd 1 ft
3 yd 2 ft —(+ 2 yd)→ 5 yd 2 ft —(+ 1 ft)→ 6 yd

⇒ 3 yd 2 ft + 2 yd 2 ft

3 yd 2 ft + 2 yd 2 ft = 5 yd 2 ft + 2 ft = 6 yd 1 ft
1 ft 1 ft

⇒ 5 yd 2 ft – 2 yd 1 ft

5 yd 2 ft – 2 yd 1 ft
5 yd 2 ft —(– 2 yd)→ 3 yd 2 ft —(– 1 ft)→ 3 yd 1 ft

⇒ 5 yd 1 ft – 2 yd 1 ft

5 yd 1 ft – 2 yd 1 ft = 3 yd 1 ft – 1 ft = 3 yd

⇒ 5 yd 1 ft – 2 yd 2 ft

5 yd 1 ft – 2 yd 2 ft = 3 yd 1 ft – 2 ft = 2 yd 2 ft
2 yd 3 ft (– 2 ft = 1 ft)

Practice

Appendix p. a8

1. (a) 6 yd 1 ft
(b) 12 yd 1 ft
(c) 20 yd 2 ft
(d) 19 yd

2. (a) 9 yd 1 ft
(b) 6 yd 2 ft

(c) 2 yd 1 ft
(d) 7 yd 2 ft
(e) 12 yd
(f) 9 yd 2 ft

3. (a) 7 yd
(b) 1 yd 2 ft

(4) Add and subtract feet and inches in compound units

Discussion

Tasks 13-15, p. 23

13: 11 inches just needs one more inch to make a foot, so we can take that from the second measurement, leaving 10 in. Tell your student that we could also add and then convert. 11 in. + 11 in. = 22 in. = 1 ft 10 in. Ask your student which method is easier.

14: We subtract from 1 ft by subtracting from 12 in.

15: We can also subtract from 1 ft when there are more feet. If we want to subtract from both feet and inches, and there are not enough inches, we can subtract from a foot instead. We then have to add back in the inches.

13. 1 ft 10 in.
14. 4 in.
15. (a) 5 ft 4 in.
(b) 5 ft 8 in.

6 ft 4 in. – 8 in. = 5 ft 8 in.
5 ft 1 ft – 8 in. = 4 in.
+ 4 in. = 8 in.

Activity

Provide your student with some more practice adding to make a 12 and subtracting from 12. You can use Mental Math 5.

Discussion

Task 16, pp. 23-24

16.(a): We can add feet first and then inches. 10 in. and 7 in. is 1 ft and 5 in., so another foot has to be added to get the final answer.

16(b): We subtract the feet first and then the inches. To subtract the inches, 7 in. are not enough to subtract 10 in. from, so we can subtract from one of the feet and add back in the 7 in.

If your student has trouble "renaming 12" mentally, you can let her rewrite the final step vertically and convert before adding and subtracting. Be sure she understands that she is converting between inches and feet, not tens and ones. So 1 ft 7 in. is 19 in. (not 17).

16. (a) 6 ft 5 in.
(b) 2 ft 9 in.

	5 ft	7 in.
+		10 in.
	5 ft	17 in.
=	6 ft	5 in.

	2	19
	~~3~~ ft	~~7~~ in.
–		10 in.
	2 ft	9 in.

Practice

Task 17, p. 24

Workbook

Exercise 8, pp. 21-22 (answers p. 26)

17. (a) 5 in.
(b) 9 in.
(c) 6 in.
(d) 5 ft 2 in.
(e) 15 in. (or 1 ft 3 in.)
(f) 2 ft 3 in.
(g) 10 yd
(h) 10 yd 2 ft

(5) Practice

Practice

Practice B, p. 25

1. (a) 15 ft (b) 263 ft (c) 925 ft
2. (a) 108 in. (b) 82 in. (c) 117 in.
3. (a) 9 yd (b) 36 yd (c) 70 yd 2 ft
4. (a) 1 ft (b) 1 ft 4 in. (c) 2 ft
5. (a) 1 ft
   (b) 2 ft
   (c) 1 ft
   (d) 5 in.
   (e) 4 in.
6. (a) 5 yd 2 ft (b) 9 yd 1 ft
   (c) 8 yd (d) 1 yd 2 ft
   (e) 8 yd 2 ft (f) 2 ft
7. (a) 12 ft 11 in. (b) 9 ft 11 in.
   (c) 12 ft 3 in. (d) 11 ft 5 in.
   (e) 9 ft 10 in. (f) 1 ft 10 in.

Reinforcement

Mental Math 6

Extra Practice, Unit 6, Exercise 3, pp. 117-118

Write down the measurements shown at the right and ask your student to put them in order from shortest to longest.

1 yd 2 ft   4 ft   39 in.   2 ft 5 in
(Answer: 2 ft 5 in, 39 in., 4 ft, 1 yd 2 ft)

1 yd 1 ft   1 m 2 cm   3 ft 10 in.   50 in.
(Answer: 1 m 2 cm, 3 ft 10 in., 1 yd 1 ft, 50 in.)

Test

Tests, Unit 6, 3A and 3B, pp. 13-20

Chapter 4 – Miles

Objectives

- Recognize the mile as a unit of measurement.
- Perform the four operations on distances in miles.

Vocabulary

- Miles

Notes

In the U.S., longer distances are measured in miles. A little over half a mile is 1 km, 6 miles is about 10 km, and 60 miles is about 100 km. (1 km equals 0.6214 miles, or 1 mile equals 1.6093 km.)

Students will learn that 1 mile = 5280 feet. 5280 is not an easy number to remember, and we do not regularly convert miles to feet in daily life. You can have your student remember that a mile is about 5000 feet and provide him with the conversion factor when he is doing workbook exercises or tests.

Since students have not yet studied 5-digit numbers or division by a 4-digit number, they will only be finding the number of feet in 1 mile and the approximate number of feet in 2 miles, and will not be converting from feet to miles.

This lesson is short, so you may want to include some of the review that follows.

(1) Understand miles as a unit of measurement

Activity

Ask your student how we measure long distances in the U.S. We use miles rather than kilometers. Discuss some distances in miles to known landmarks or between known landmarks. You can draw a roughly scaled map and mark the distances in miles.

Tell your student that a mile is a little longer than one and a half kilometers. 10 kilometers is about 6 miles. 100 kilometers is about 60 miles.

You may want to tell your student the following facts, and have your student look up other distances.

⇒ The distance around the earth at the equator is about 25,000 miles

⇒ The highest mountain is about 5 miles high.

⇒ The deepest trench (in the ocean) is about 10 miles deep.

⇒ The distance from the North Pole to the South Pole through the center of the earth is about 8,000 miles.

Discussion

Concept page 26

Point out that the abbreviation for a mile is mi. 1 mile is 5280 feet. Ask your student to find the number of yards in a mile.

1 mile = 5280 feet

5280 ÷ 3 = 1760

1 mile = 1760 yards

Tasks 1-2, p. 26

1. 1 day: 3 mi
   7 days: 3 mi x 7 = 21 mi
2. **2930 mi**
   2930 mi – 1260 mi = 1670 mi
   The distance between New York and Denver is **1670 mi.**

Workbook

Exercise 9, p. 23 (answers p. 26)

Reinforcement

Have your student solve the following problems.

⇒ It is 1390 miles when you fly from Boston to Minneapolis and 1650 miles when you fly from Minneapolis to Seattle.

(a) How far is it from Boston to Seattle if the flight stops over in Minneapolis?
(1390 mi + 1650 mi = 3040 mi)

(b) How much further is it from Minneapolis to Seattle than from Minneapolis to Boston?
(1650 mi – 1390 mi = 260 mi)

⇒ Put in order from shortest to longest.

1500 yd   1 mi   2 km   5000 ft   (1500 yd, 5000 ft, 1 mi, 2 km)

Extra Practice, Unit 6, Exercise 4, pp. 119-120

Test

Tests, Unit 6, 4A and 4B, pp. 21-28

Review 6

Review

Review 6, pp. 27-29

Reviews are cumulative, so this review covers material from *Primary Mathematics* 3A as well.

Your student should be allowed to choose when to apply mental math strategies, which can vary from problem to problem, or the standard algorithms, which can be done with any type of computation problem.

For problem 24, you can tell your student that there are 5280 feet in a mile.

Workbook

Review 6 pp. 24-27 (answers p. 26)

Test

Tests, Units 1-6, Cumulative A and B, pp. 29-40

1. (a) 93 (b) 100 (c) 102
2. (a) 198 (b) 197 (c) 72
3. (a) 33 (b) 34 (c) 5
4. (a) 1 (b) 8 (c) 1
5. (a) 120 (b) 400 (c) 1800
6. (a) 100 (b) 60 (c) 200
7. (a) 600 (b) 450 (c) 2100
8. (a) 20 (b) 80 (c) 90
9. (a) 600 (b) 240 (c) 2500
10. (a) 55
 (b) 242
 (c) 510
 (d) 5494
11. 1 notebook: 80 pages
 6 notebooks: 80 x 6 = **480** pages
12. 5 bags: 200 onions
 1 bag: 200 ÷ 5 = **40** onions
13. Friday (70), Sunday (?)
 70 x 4 = 280
 He sold **280** muffins on Sunday.

14. (a) 2 (b) 465 – 456 = **9**
 (c) 34 (c) 350 – 251 = **99**
15. (a) 6372 (b) 3664 (c) 891
 (d) 27 R 3 (e) 1786 R 2 (f) 211 R 3
16. (a)

	Boys	Girls
Walkathon	56	50
Swimming	63	66
Carnival	45	47

 (b) 327
17. (a) 2
 (b) 4
18. (a) 395 cm (b) 405 cm
 (d) 2060 m (d) 3078 m
 (e) 12 ft (f) 38 ft
 (g) 72 in. (h) 43 in.
19. (a) 6 m 18 cm (b) 9 m 36 cm
 (c) 3 km 90 m (d) 3 km 999 m
 (e) 11 yd 0 ft (f) 651 yd 1 ft
20. (a) 45 cm (b) 1 m 5 cm
 (c) 400 m (d) 11 km 725 m
 (e) 3 m 91 cm (f) 1 km 50 m
 (g) 3 yd 1 ft (h) 5 ft 6 in.
21. 1 m 80 cm + 1 m 65 cm = 3 m 45 cm
 The total length is **3 m 45 cm**.
22. 4 km 400 m – 2 km 940 m = 1 km 460 m
 The distance between the boat and the lighthouse is **1 km 460 m**.
23. 6 ft 2 in. – 3 ft 10 in. = 2 ft 4 in.
 Josh is **2 ft 4 in.** shorter than his Dad.
24. 5280 ft ÷ 3 ft/yd = 1760 yd
 There are **1760 yards** in a mile.

Workbook

Exercise 1, p. 7

1. (a) 200 cm (b) 300 cm
   (c) 500 cm (d) 900 cm

2. (a) 4 m (b) 6 m
   (c) 7 m (d) 8 m

3. (a) 150 cm
   (b) 328 cm
   (c) 509 cm

4. (a) 2 m 10 cm
   (b) 2 m 75 cm
   (c) 2 m 6 cm

5. (a) <
   (b) =
   (c) >

Exercise 2, p. 8

1. (a) 47
   (b) 15
   (c) 26
   (d) 22
   (e) 3
   (f) 38
   (g) 68
   (h) 56

2. (a) 10
   (b) 35
   (c) 95
   (d) 70
   (e) 47
   (f) 18
   (g) 93
   (h) 27

Exercise 3, pp. 9-10

1. (a) 3 m 85 cm
   (b) 4 m 70 cm
   (c) 6 m 10 cm

2. (a) 4 m 20 cm
   (b) 5 m 85 cm
   (c) 7 m 68 cm
   (d) 4 m 26 cm
   (e) 7 m 18 cm

3. (a) 1 m 10 cm
   (b) 2 m 59 cm
   (c) 6 m 39 cm

4. (a) 1 m 89 cm
   (b) 4 m 12 cm
   (c) 3 m 85 cm
   (d) 3 m 86 cm

Exercise 4, pp. 11-14

1. (a) 582 km + 180 km = **762 km**
   (b) San Francisco
   582 km – 120 km = **462 km**

2. (a) 23 km
   (b) 90 km
   (c) 65 km
   (d) 19 km
   (e) 6 km

3. (a) 2000 (b) 4000
   (c) 5000 (d) 8000

4. (a) 3 (b) 6
   (c) 7 (d) 9

5. (a) 1145
   (b) 3050
   (c) 1298
   (d) 2078
   (e) 2580
   (f) 1006
   (g) 3670

6. (a) 1 km 732 m
   (b) 1 km 305 m
   (c) 2 km 245 m
   (d) 1 km 300 m
   (e) 3 km 260 m
   (f) 3 km 6 m
   (g) 2 km 108 m

7. (a) > (b) > (c) <

8. (a) 741 m
   (b) 995 m + 870 m = **1865 m = 1 km 865 m**
   (c) well
   1865 m – 741 m = **1124 m = 1 km 124 m**
   (d) 968 m + 968 m = **1936 m**
   (e) 865 m + 995 m + 741 m = **2 km 601 m**

Workbook

Exercise 5, pp. 15-16

1.

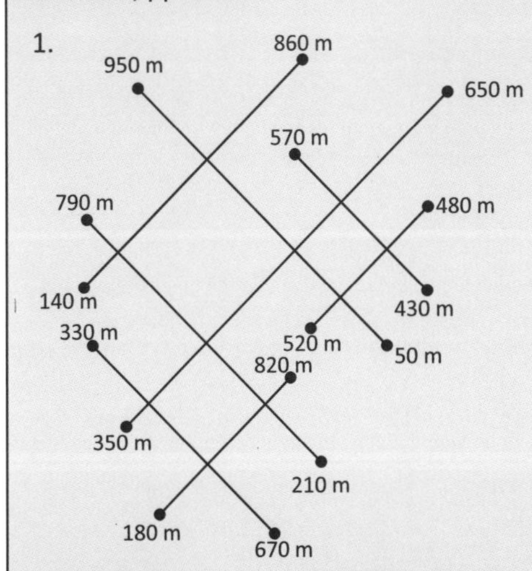

2. (a) 20 m (b) 110 m
   (c) 210 m (d) 580 m
   (e) 80 m (f) 120 m

3. (a) 510 m (b) 980 m
   (c) 1993 m (d) 550 m
   (e) 2985 m
   (f) 1 km 810 m
   (g) 6 km 173 m
   (h) 9 km 282 m

4. 1 km – 450 m = 550 m.
   She has **550 m** to go.

Exercise 6, pp. 17-18

1. (a) 1 km 850 m
   (b) 3 km 180 m
   (c) 5 km 230 m

2. (a) 6 km 110 m
   (b) 7 km 970 m
   (c) 10 km 200 m
   (d) 6 km 150 m
   (e) 9 km 200 m
   (f) 11 km 100 m

3. (a) 2 km 70 m
   (b) 3 km 940 m
   (c) 5 km 260 m

4. (a) 2 km 650 m
   (b) 5 km 920 m
   (c) 3 km 750 m
   (d) 6 km 920 m

Exercise 7, pp. 19-20

1. (a) 29 ft
   (b) 10 ft
   (c) 381 ft
   (d) 602 ft

2. (a) 5 yd 0 ft
   (b) 8 yd 1 ft
   (c) 101 yd 2 ft
   (d) 200 yd 0 ft

3. (a) 23 in.
   (b) 101 in.
   (c) 122 in.
   (d) 34 in.
   (e) 81 in.
   (f) 108 in.

4. (a) 15 in. = 15 in. (b) 5 ft < 6 ft
   (c) 71 in. > 70 in. (d) 32 ft > 5 ft
   (e) 16 ft = 16 ft (f) 22 in. < 26 in.

5. (a) 450 ft + 361 ft = **811 ft** or **270 yd 1 ft**
   (b) 678 ft + 107 ft = **785 ft** or **261 yd 2 ft**
   (c) **D**; **22 ft** or **7 yd 1 ft**
   43 yd = 129 ft;
   129 ft – 107 ft = 22 ft
   or
   107 ft = 35 yd 2 ft;
   43 yd – 35 yd 2 ft = 7 yd 1 ft
   (d) 450 ft + 678 ft = **1128 ft**

Workbook

Exercise 8, pp. 21-22

1. (a) 2 ft 4 in.
   (b) 3 ft 7 in.
   (c) 6 ft 7 in.
   (d) 4 yd 1 ft

2. (a) 5 yd 1 ft
   (b) 9 yd 0 ft
   (c) 21 yd 1 ft
   (d) 11 ft 5 in.
   (e) 13 ft 0 in.
   (f) 10 ft 4 in.

3. (a) 2 ft 2 in.
   (b) 4 ft 9 in.
   (c) 7 ft 11 in.
   (d) 5 yd 0 ft
   (e) 1 yd 2 ft

4. (a) 1 yd 2 ft
   (b) 1 yd 0 ft
   (c) 1 yd 2 ft
   (d) 0 ft 4 in.
   (e) 4 ft 8 in.
   (f) 0 ft 10 in.

5. (a) C
   (b) 0 ft 7 in.
   (c) 19 ft 6 in.

Exercise 9, p. 23

1. (a) >
   (b) >

2. (a) 3806 mi + 5950 mi = **9756 mi**
   (b) 5950 mi – 3806 mi = **2144 mi**

3. 5280 ft/mi, so **Madeline** lives closer.

4. (a) mi
   (b) ft
   (c) yd
   (d) in.
   (e) mi

Review 6, pp. 24-27

1. odd: 9, 13, 127, 1229
   Even: 6, 72, 354, 1350

2.

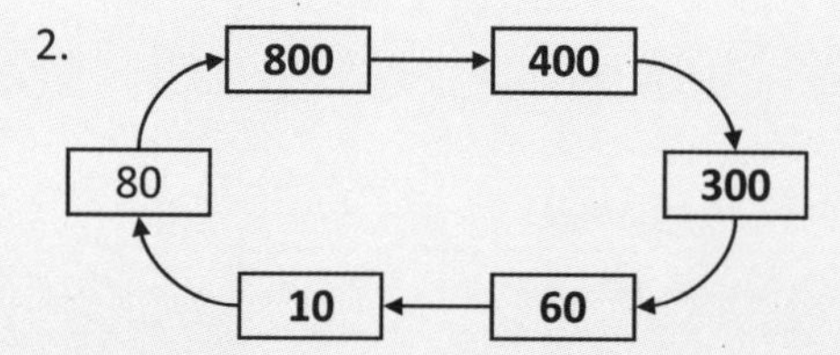

3. (a) 6250
   (b) 5403
   (c) 5000
   (d) 2009

4. (a) 8 cm
   (b) 45 cm
   (c) 3 in.
   (d) 930 m
   (e) 310 m
   (f) 1 ft
   (g) 40 cm
   (h) 44 cm
   (i) 1 m 29 cm

5. 3 km 120 m – 1 km 250 m = **1 km 870 m**

6. (a) Thursday
   (b) 210 + 150 + 90 + 60 + 120 = **630** eggs

7. (a) likely
   (b) unlikely
   (c) certain
   (d) impossible

8. 5 books: $90
   1 book: $90 ÷ 5 = $18
   1 book costs **$18.**

9.

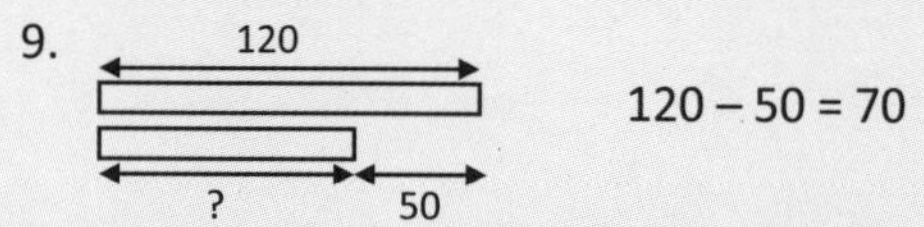

   The smaller number is **70.**

10. How many 8's in 72?
    72 ÷ 8 = 9
    The length is **9** times as long as the width.

Unit 7 – Weight

Chapter 1 – Kilograms and Grams

Objectives

- Estimate and weigh in kilograms and grams.
- Convert between kilograms and grams.
- Add and subtract kilograms and grams in compound units.

Vocabulary

- Kilogram
- Gram

Notes

Students weighed objects and read scales in either kilograms *or* grams in *Primary Mathematics* 2A. In this chapter they will weigh objects in compound units, kilograms *and* grams. Students will also learn to convert between kilograms and grams and to add and subtract in compound units. Since 1 kilogram equals 1000 grams, the process is the similar to adding and subtracting kilometers and meters.

If possible, give your student practical work in measuring weight, and in using different weighing scales, particularly if he has not used earlier levels of *Primary Mathematics*. Encourage him to estimate a weight before measuring it. If you do not have kilogram or gram weights, you can substitute other items. It is not necessary at this stage to use very accurate weights. Sometimes the bulk section in a supermarket will have scales that also show kilograms, and you can weigh a bag of beans or other item to use as your kilogram weight. Modeling clay is sometimes sold in one pound blocks; two and a fifth of these is a about a kilogram. A quart (or liter) of water weighs about 1 kg. Four hundred pennies weigh about 1 kg. The unit cube from a base-10 set weighs about 1 gram, or two regular paper clips weigh about a gram.

Strictly speaking, the kilogram and gram are units of mass (a measure of the amount of matter in the object), not of weight (a measure of the force of gravity on an object). However, since we use the term weight in daily speech when weighing things, it will be used here.

If your student is familiar with his own weight, it may seem odd that the children in the word problems only weigh perhaps 35 kilograms, when he weighs maybe about 60 pounds. Since 1 kilogram weighs 2.2 pounds, that is, a bit over two pounds, you can double the weight in kilograms to get an estimate of the weight in pounds. So when the textbook says a person weighs 35 kilograms, then that person weighs about 70 pounds (and is likely a child). An adult would weigh about 60 kilograms or more.

Material

- Kilogram and gram weights
- Balance or weighing scale in kilograms and grams

(1) Weigh objects in kilograms and grams

Discussion

Concept p. 30

800 g	1 kg 300 g

Tell your student that the kilogram and gram are part of the metric system of measurement. Show her a kilogram weight and a gram weight and let her handle them and feel the difference. Point out the abbreviations kilograms (kg) and grams (g). Tell her that 1 kilogram is 1000 grams.

Have your student read the weights on the scales on this page. Make sure she understands what each of the intervals between the tick marks represent.

Activity

If you have a scale, let your student weigh some items in kilograms and grams. Then get him to estimate the weights first, similar to task 2 on p. 31 of the textbook.

Discussion

Task 1, p. 31

1. (a) 1 kg	(b) 1 kg 200 g
(c) 900 g	(d) 1 kg 700 g

After your student has written down the weights, ask her why the pointer goes farther around in 1(c) than in 1(a), but the weight is less. Make sure she understands that the scales have been calibrated differently; in the top scales one full turn is 4 kg whereas in the bottom scales one full turn is 2 kg.

Workbook

Exercise 1, pp. 28-29 (answers p. 41)

Enrichment

Write the following problem and discuss ways to solve it with your student. A possible method of solution is given here.

⇒ There are three postcards, A, B, and C. A and B together weigh 45 g. B and C together weigh 41 g. All three postcards together weigh 70 g. How much does each postcard weigh?

Get your student to diagram the information. As he figures out some weights, he can label the postcards with those weights.

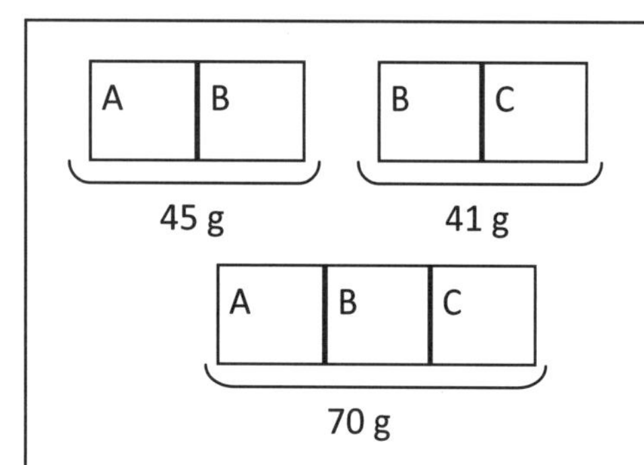

Difference between A and C: AB – BC = 45 g – 41 g = 4 g

Weight of A: ABC – BC = 70 g – 41 g = 29 g

Weight of C: 29 g – 4 g = 25 g

Weight of B: 41 g – 25 g = 16 g

Check: A + B + C = 29 g + 16 g + 25 g = 70 g

(2) Convert between kilograms and grams

Activity

Since 1 kg = 1000 g, the processes used here are the same as for converting between kilometers and meters, where 1 km = 1000 m. Your student will probably be able to do the learning tasks independently. You can discuss the following problems first.

⇒ 4 kg = ______ g	4 kg x 1000 g/kg = **4000** g
⇒ 4 kg 345 g = ________ g	4 kg 345 g = 4000 g + 345 g = **4345** g
⇒ 4 kg 5 g = ________ g	4 kg 5 g = 4000 g + 5 g = **4005** g
⇒ 2000 g = _____ kg	Each thousand grams is 1 kg. 2000 g = **2** kg
⇒ 10,000 g = _____ kg	10,000 g = **10** kg
⇒ 2345 g = _____kg ______ g	2345 g = **2** kg **345** g / \\ 2000 g 345 g
⇒ 5002 g = _____ kg _____ g	5002 g = **5** kg **2** g / \\ 5000 g 2 g

Practice

Tasks 3-7, p. 32

Workbook

Exercise 2, pp. 30-33 (answers p. 41)

3. 2200 g
4. (a) 1456 g (b) 2370 g (c) 3808 g
   (d) 2080 g (e) 1008 g (f) 4007 g
5. 350 g x 4 = 1400 g = **1** kg **400** g
6. (a) 2 kg 143 g (b) 1 kg 354 g (c) 3 kg 800 g
   (d) 2 kg 206 g (e) 3 kg 85 g (f) 4 kg 9 g
7. The **chicken** is heavier by **150 g**.
   (1250 g – 1100 g = 150 g
   Or 250 g – 100 g = 150 g)

(3) Add and subtract kilograms and grams

Discussion

Task 8, p. 33

We can use the mental math strategy of "subtract from 1000" or use the subtraction algorithm by converting the measurements to grams, subtracting, and converting the answer back to kilograms and grams.

8. 150 g
   150 g

1000 – 850 = 150
2 kg – 1 kg 850 g = 150 g
/ \
1 kg   1 kg – 850 g
– 1 kg
Or
  2 0 0 0
– 1 8 5 0
    1 5 0 → 150 g

Practice

Task 9, p. 33

9. (a) 605 g
   (b) 915 g
   (c) 600 g
   (d) 940 g
   (e) 460 g
   (f) 2 kg 195 g

Discussion

Task 10, p. 34

Discuss strategies for adding and subtracting in compound units. We can add or subtract kilograms first, and then add or subtract grams using mental math strategies or the standard algorithms for addition or subtraction.

10. (a) 5 kg 40 g
    (b) 1 kg 120 g

Practice

Tasks 11-14, p. 35

Workbook

Exercise 3, pp. 34-36 (answers p. 41)

Reinforcement

Extra Practice, Unit 7, Exercise 1, pp. 127-130

Test

Tests, Unit 7, 1A and 1B, pp. 41-47

11. (a) 2 kg 600 g + 1 kg 500 g = 4 kg 100 g
    The total weight is **4 kg 100 g**.
    (b) 2 kg 600 g – 1 kg 500 g = 1 kg 100 g
    The difference in weight is **1 kg 100 g**.

12. (a) 5 kg 500 g (b) 5 kg 100 g
    (c) 5 kg (d) 6 kg 120 g
    (e) 2 kg 810 g (f) 3 kg 250 g
    (g) 2 kg 750 g (h) 2 kg 95 g

13. 32 kg – 25 kg 750 g = 6 kg 250 g
    She gained **6 kg 250 g**.

14. (a) 2 kg 990 g + 4 kg 200 g = 7 kg 190 g
    The total weight is **7 kg 190 g**.
    (b) 4 kg 200 g – 2 kg 990 g = 1 kg 210 g
    The difference in weight is **1 kg 210 g**.

Chapter 2 – Word Problems

Objectives

- Solve word problems involving measurement.

Notes

In *Primary Mathematics* 3A, students were introduced to bar models as a way to diagram the information in a word problem in order to help them determine what equation to use, or what the first step would be in a two-step solution. There are two basic types of models, a part-whole model and a comparison model. These then can be combined or modified depending on the problem.

A part-whole model is generally used when the problem requires finding the whole or a missing part. For example:

⇒ There are 106 blue and red marbles. 69 of them are red. How many are blue?

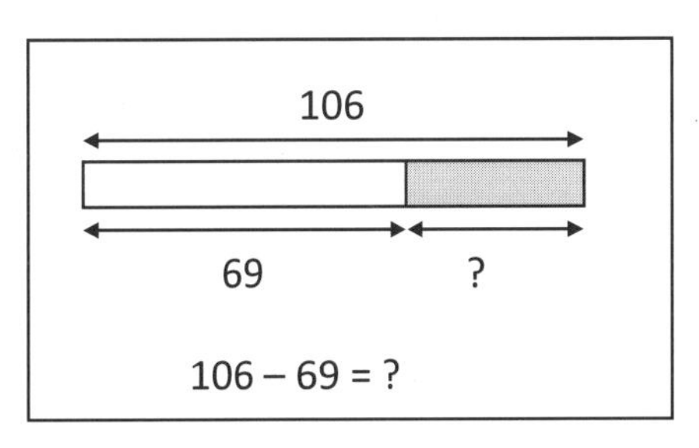

From the model, the student can see that we need to subtract to find the answer.

A comparison model is generally used when we are comparing two or more numbers. For example:

⇒ There are 302 red marbles. There are 128 more blue marbles than red marbles. How many blue marbles are there?

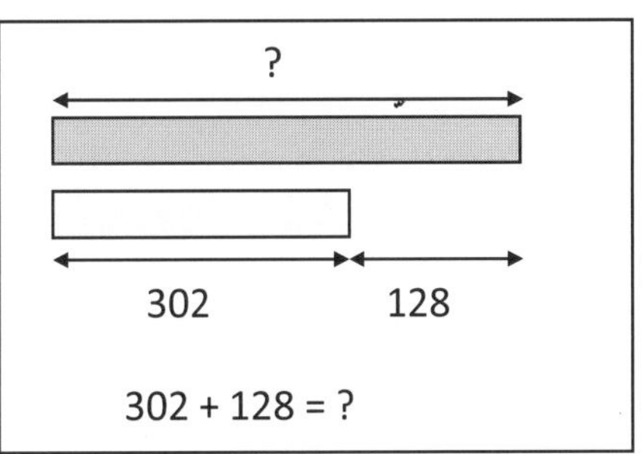

We can draw two bars, one for each type of marble. From the problem, we know that there are more blue marbles than red marbles, so we draw that bar longer. From the model, we can see that we need to add to find the number of blue marbles.

For multiplication and division, we can also use part-whole and comparison models, but now we draw equal parts. These equal parts are called units, and in solving the problems, we generally need to find the value of 1 unit. For example:

⇒ 432 marbles are divided into 4 bags. How many marbles are in each bag?

A part-whole model would help students see that division is needed to solve the word problem.

⇒ These 4 times as many blue marbles as red marbles. There are 62 red marbles. How many blue marbles are there?"

Since we are comparing two quantities, we can use a comparison model.

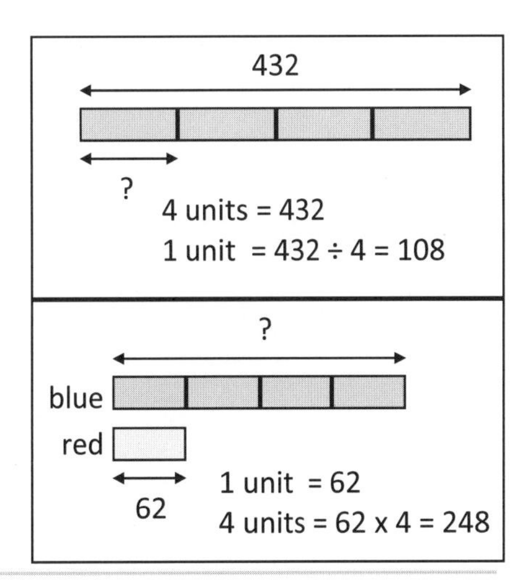

These examples illustrating the basic models are fairly simple, and students have been doing problems like them without the use of bar models. Bar models are a more powerful tool when used with problems that involve more than one step and a combination of operations. For example

⇒ There are 3 times more blue marbles than red marbles. There are 555 marbles in all. How many blue marbles are there?

The model makes it easier to see that if there are three times more blue than red marbles, then there must be a total of 5 units, and that we need to find the value of 4 units to answer the problem.

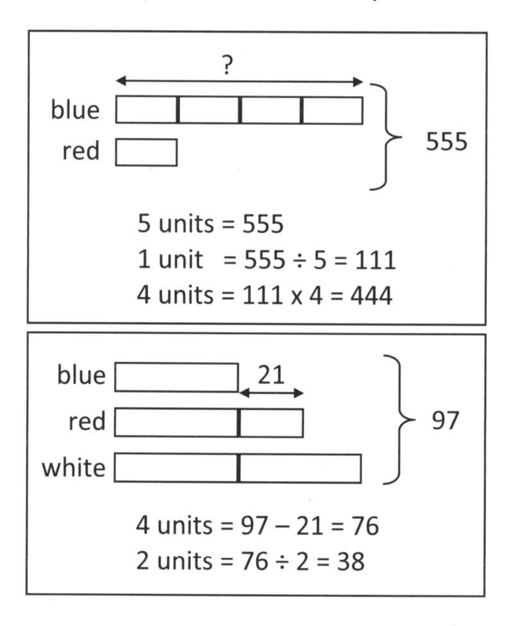

⇒ There are 97 marbles. There are 21 more red marbles than blue marbles. There are twice as many white marbles as blue marbles. How many white marbles are there?

A model shows us that we can get 4 equal units by subtracting 21 from the total. Since we want to find the value of 2 units, we simply have to find half of 4 units. Or we can find the value of one unit, and then multiply by 2 to get the value of 2 units.

Bar models are a problem solving tool. Do not require your student to draw models if he can solve the problem without them. However, he should be capable of using bar models when needed, so any time he gets an answer wrong, require him to draw a bar model, guiding him as needed. If he consistently gets problems wrong, you may have to require models for a while for all problems.

Do not impose a set of arbitrary steps on drawing bar models; there are none that will apply to all types of problems your student will encounter, and imposing a set of specific steps short-changes development of problem solving skills and the ability to use logic in solving these problems. It is not sufficient, for example, to tell your student to draw a unit for each thing the problem is about, and then add units or parts or labels while going sentence by sentence through the problem. If the problem does not fit those steps, either the steps will not work, or a lot of back-tracking will have to be done. Drawing a model starting with some part of the problem other than the first sentence might result in a simpler model that is easier to interpret.

In the textbook, models are shown for most of the problems in the learning tasks for this chapter. It takes more skill to determine how to model the problem than it does seeing a model that is already drawn and determining what equation to use. You may want to write down the problems and work through the steps in drawing the models with your student rather than just having her look at the completed models. Make sure she reads the whole problem before trying to draw a model.

If your student has done previous levels in *Primary Mathematics*, this chapter is primarily review. If he has not done previous levels in *Primary Mathematics*, you may want to spend more time on this lesson.

(1) Solve word problems

Discussion

Concept page 36

850 g

This is a part-whole type of problem, with the weight of the marbles the missing part. The model shows two parts, with the total and one part labeled. From this, we can see that we can solve the problem with subtraction. Point out that the text uses curly braces, but if those are too hard to draw, we can use simple arrows instead.

Tasks 1-4, pp. 37-38

1. 560 g – 305 g = 255 g
   There are **255 g** of sauce.
2. 400 g
   400 g
3. 19 kg
   19 kg
4. 5 units = 950 g x 5
   = 4750 g.
   = **4 kg 750 g**

1: You may want to ask your student to draw a model for this problem.

bottle of sauce: 560 g
empty bottle: 305 g ?

2: There are two parts, the football and the tennis balls. We are given the total weight, and we need to find the weight of the football. Instead of being given the weight of all the tennis balls, we are only given the weight of one of them. Since all the balls have the same weight, we can divide the part for the tennis balls up into 10 units and label one of them as 60 g. The first step, then, is to find the weight of all the tennis balls, and then we can subtract to find the weight of the football.

3: In this problem we are comparing the weight of two children, and want to find the weight of one of them. When we compare numbers, we can align the bars for each at the left so that we can see and label the difference between the bars, which represents the difference between the numbers. Point that whenever we are given how many times one part is than another, as in the second sentence, that is a good place to start in drawing a model. So we would first draw 3 bars for William and one for Sean, and then label them with what we know and what we need to find. Also have your student find the following:

⇒ How heavy are both children? Discuss 2 methods. In the first method, we use the value of 1 unit to find the value of 4 units. In the second, we add the two weights.

1) 1 unit = 19 kg
   4 units = 19 kg x 4 = 76 kg
2) 57 kg + 19 kg = 76 kg

⇒ How much heavier is William than Sean? Discuss 3 methods. Again, in the first method we can use the value of 1 unit to find the value of 2 units. In the second method, we subtract the weights.

1) 2 units = 19 kg x 2 = 38 kg
2) 57 kg – 19 kg = 38 kg
3) If we have already found 4 units, then 2 units is half that:
   2 units = 76 kg ÷ 2 = 38 kg

Point out that once we find the value of 1 unit, it is easy to answer different types of questions.

4: Remind your student that we always want to convert measurements for the final answer, if possible.

Tasks 5-7, pp. 38-39

5. 33 kg 800 g

6. (a) 3 kg 200 g + 1 kg 800 g = 5 kg
The total weight is **5 kg**.
(b) 3 kg 200 g – 1 kg 800 g = 1 kg 400 g
The difference in weight is **1 kg 400 g**.

7. (a) 4 kg
(b) 5 kg 50 g – 4 kg = **1 kg 50 g**

34 kg 600 g
J
D
?
800 g

5: Note how small 800 g is compared to 34 kg 600 g. Tell your student that it is not necessary to be absolutely precise in the relative sizes of the bars, or to draw the bars perfectly; the purpose of the models is to give us an idea of how to solve the problem. We do know from the problem that John's bar needs to be longer, so we do have to draw that bar longer. Rather than rectangles for bars, we could even use lines with little bars at the end. Even if we draw a simple model, it is good to get into the habit of labeling the bars when we have two or more of them, even if we just use initials. When problems get more complicated, or when we are comparing more than two things, labeling the bars with names or initials will help us keep track of what each one is for.

6: This model shows us that we can draw one model to show both things we want to find. For 6(a) we could have instead shown two bars side by side, since there are two parts and we want to find a whole, but in 6(b) the problem is asking for the difference in weight, so we would draw two bars, one on top of each other. We can label the total using a brace on the side. Your student does not have to use a curly brace; they are sometimes hard to draw.

7: You may want to have your student draw a model. You can use this problem as an opportunity to show that several kinds of models could be drawn to show the same information.

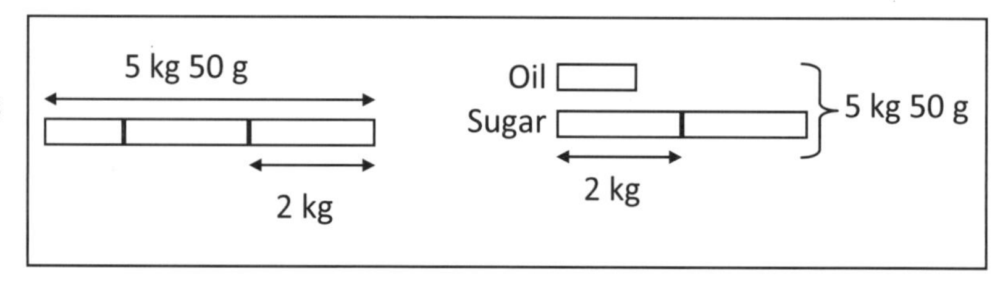

Tasks 8-10, p. 39

These last three problems deal with finding a unit price. It is not necessary to draw models for these, but it can help to visualize them, or write what the total weight is first, as shown in the answers here, so that it is easy to see what we need to divide by. To go from 5 kg to 1 kg, we divide by 5, so we also divide the cost by 5. If needed, you can draw bar models, where each kilogram is 1 unit. We know the total cost, so to find the cost of each unit we divide by the number of units.

8. (a) 5 kg
(b) 5 kg: $5
1 kg: $5 ÷ 5 = **$1**

9. (a) 2 kg
2 kg: $6
1 kg: $6 ÷ 2 = **$3**

10. 8 kg: $24
1 kg: $24 ÷ 8 = $3
1 kg of nails cost **$3**.

Workbook

Exercise 4, pp. 37-39 (answers p. 42)

Reinforcement

Extra Practice, Unit 7, Exercise 2, pp. 131-132

(2) Practice

Practice

Practice A, p. 40

Encourage your student to draw bar models for problems 6 and 7.

6: The second sentence tells us that Brian is 5 times as heavy as his son, so draw two bars on top of each other, one with 5 units and one with 1 unit. Then label with the information in the problem.

7: The second sentence tells us that Hugh is twice as heavy as David, so draw that first. Then draw a bar for Matthew which is shorter than Hugh's bar and mark the difference 27 g. Label Matthew's bar with a question mark. From the diagram, we can see that we need to first find Hugh's weight.

8: Your student has not yet learned how to divide 5 by 10. Encourage her to come up with a way to solve this based on her knowledge of money, and the concept of converting units. We can convert the dollars to cents, and then it is easy to divide 500 cents by 10.

1. (a) 5000 g (b) 1950 g (c) 1060 g
   (d) 2805 g (e) 2005 g (f) 3002 g
2. (a) 1 kg 905 g (b) 1 kg 55 g (c) 2 kg 208 g
   (d) 3 kg 390 g (e) 3 kg 599 g (f) 5 kg 2 g
3. (a) 3 kg 240 g (b) 5 kg 100 g
   (c) 2 kg 520 g (d) 2 kg 570 g
4. (a) 2 kg 50 g + 3 kg 20 g = 5 kg 70 g
   The total weight is **5 kg 70 g**.
   (b) 3 kg 20 g – 2 kg 50 g = 970 g
   The difference in weight is **970 g**.
5. 100 kg – 46 kg 540 g = 53 kg 460 g
   Sam's weight is 53 kg 460 g.
6.

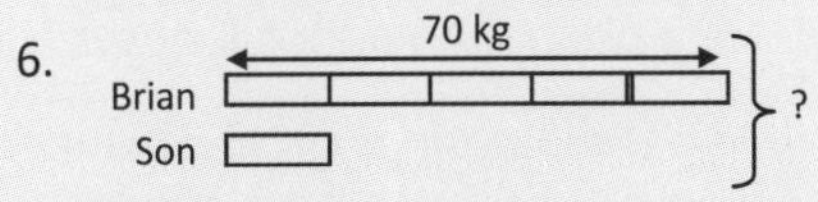

5 units = 70 kg
1 unit = 70 kg ÷ 5 = 14 kg
6 units = 14 kg x 6 = 84 kg
Or: 70 kg + 14 kg = 84 kg
Their total weight is **84 kg**.

7.

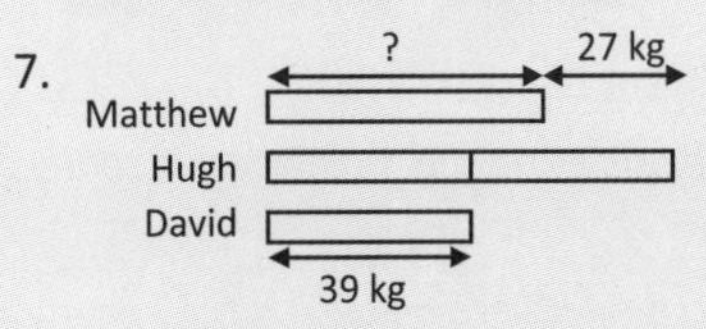

Hugh's weight: 39 kg x 2 = 78 kg
Matthew's weight: 78 kg – 27 kg = 51 kg
Matthew weighs **51 kg**.

8. 10 kg: $5 = 500¢
   1 kg: 500¢ ÷ 10 = 50¢
   1 kg of apples costs **50¢**.

Tests

Tests, Unit 7, 2A and 2B, pp. 49-55

Enrichment

⇒ The total length of 4 strings is 100 cm. String A is 9 cm shorter than string B. String B is three times as long as String C. String D is 13 cm longer than String C. How long is the longest string?

Since we are told that string B is three times as long as String A, draw strings B and C first. We do not know at first how much shorter String A is than B, or how much longer String D is than B, or really which is the longest string, but how we draw A and D will not change the method used to solve this problem. If we add 9 cm to A, and subtract 13 cm from D, then we will have 8 units. After we find that 1 unit is 12 cm, we know that B is the longest string, even if our initial drawing is not exact.

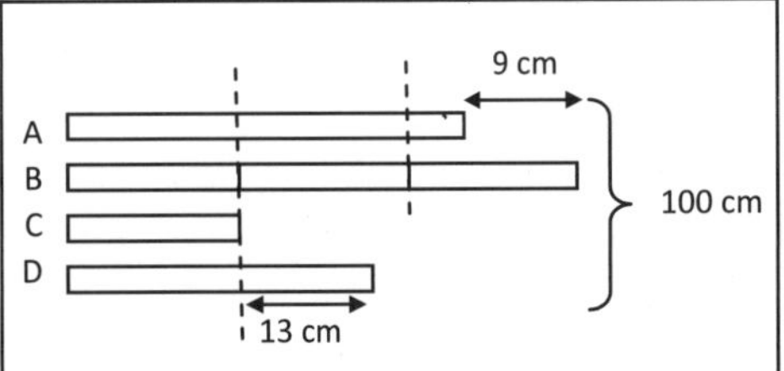

8 units = 100 cm + 9 cm – 13 cm = 96 cm
1 unit = 96 cm ÷ 8 = 12 cm
3 units = 12 cm x 3 = 36 cm
The longest string is 36 cm long.

Chapter 3 – Pounds and Ounces

Objectives

- Measure weight in pounds and ounces.
- Compare kilogram to pound and gram to ounce.
- Convert between pounds and ounces.
- Add and subtract weight in pounds and ounces.

Vocabulary

- Pounds
- Ounces

Notes

Pounds and ounces are customary units of weight in the US. They are a measure of weight, not mass. Your student may be more familiar with pounds than with kilograms. A kilogram is about twice as heavy as a pound. An ounce is about 30 times heavier than a gram.

On earth, 1 kg = 2.205 pounds or 35.28 ounces, and 1 ounce = 28.35 grams.

Give your student practical work in estimating and measuring weights so that she can make reasonable estimates. If you do not have pound or ounce weights, you can substitute other items. It is not necessary at this stage to use very accurate weights. Beans or other items are often sold in pound bags. About 60 paper clips, 11 pennies, or 5 quarters weigh about one ounce. You can tape together 5 quarters to be your ounce weight.

The abbreviation for pound, **lb**, comes from the Latin word *libra*, a unit of weight. The abbreviation for ounce, **oz**, comes from the 15th century Italian word *onza*.

Since students have not yet learned to divide by a 2-digit number, any conversions from ounces to pounds will involve only small multiples of 16, so that students can find the number of pounds and left-over ounces using repeated subtraction or counting up by 16's.

Material

- Pound and ounce weights
- Balance or weighing scale in pounds and ounces
- Kilogram weight and gram weight
- Mental Math 7 (Appendix)

(1) Convert between pounds and ounces

Activity

Show your student a pound weight and an ounce weight and let her hold them to feel their heft. Tell her that we often weigh things in pounds and ounces, but these units of weight are not used in science or in most of the world anymore.

Show your student a kilogram and a gram weight and let him feel the difference with the pound and ounce weight. If you have a balance, he can see about how many pounds are in a kilogram (about 2) and grams in an ounce (about 28). A kilogram is heavier than an pound, and an ounce is quite a bit heavier than a gram.

You can also have your student use the balance to see how many ounces make up a pound (16).

Discussion

Concept p. 41

> 8 oz
> 1 lb 11 oz

Point out the abbreviation for a pound (lb) and for an ounce (oz) and that 1 lb is the same as 16 ounces. As with kilograms and grams, we can weigh items in pounds and ounces. Point out the expanded scale at the top of p. 41. With scales in the metric system, the divisions are usually in tens. But in this case, there are 16 divisions between 0 lb and 1 lb. A longer line at the half-way mark is therefore 8 (not 5). Have your student tell you what the longer marks are for in the scale at the bottom right of p. 41.

Task 1 p. 42

> 1. (a) 2 lb (b) 7 lb 4 oz

Help your student interpret the scales. Each tick mark is 2 oz.

Activity

If you have a scale or balance, have your student estimate the weight of various items and weigh them. You can draw a chart with the name of the item, the estimated weight, and the measured weight. Estimating weights and then measuring them will help your student become more familiar with the weights. In estimating weights, it is helpful to have in mind the weight of a familiar objects. For example, 4 apples are about a pound, so 1 apple is about 4 ounces. Generally we weigh things like mushrooms and berries, which are fairly expensive, in ounces. A basket of berries at the farmer's market will usually be less than a pound, for example. It is unlikely that she will ever have to estimate in ounces.

Write 1 lb = 16 ounces. Ask your student how we can convert from pounds to ounces. We multiply the number of pounds by 16 oz/lb. Ask her to find the number of ounces for 2 and 3 pounds. You may want to continue to 10 pounds. She can try to find the answer mentally or use the standard multiplication algorithm. It is not necessary for her to memorize the multiplication facts for 16.

> 1 lb = 16 oz
> 2 lb x 16 oz/lb = 32 oz
> 3 lb x 16 oz/lb = 48 oz
> 4 lb x 16 oz/lb = 64 oz
> 5 lb x 16 oz/lb = 80 oz
> 6 lb x 16 oz/lb = 96 oz
> 7 lb x 16 oz/lb = 112 oz
> 8 lb x 16 oz/lb = 128 oz
> 9 lb x 16 oz/lb = 144 oz
> 10 lb = 16 oz x 10 = 160 oz

Discussion

Tasks 2-8, pp. 42-43

2-3: To convert from pounds and ounces to ounces, we convert the pounds to ounces by multiplying them by 16, and then add the rest of the ounces.

4-5: These are all between 1 and 2 pounds, so the answers are simply 1 pound and the difference between the number of ounces given and 16.

6: To subtract from a pound, we subtract ounces from 16. Ask your student to subtract other amounts of up to 15 ounces from a pound.

7: To subtract from several pounds, we can simply subtract from one of the pounds.

8: To add ounces, we can add by "making 16" as shown in this task, or we can add the ounces and then convert by subtracting the answer from 16 to make a pound. Give your student some other examples.

2. 4 lb 13 oz = 64 oz + 13 oz = **77 oz**

3. (a) 5 lb = 5 lb x 16 oz/lb = **80 oz**
   (b) 7 lb 15 oz = 112 oz + 15 oz = **127 oz**
   (c) 9 lb 9 oz = 144 lb + 9 oz = **153 oz**

4. 8 oz x 3 = 24 oz = **1 lb 8 oz**

5. (a) 1 lb (b) 1 lb 4 oz (c) 1 lb 10 oz

6. 2 oz

7. **6** oz
   2 lb **6** oz

8. **4** oz
   1 lb **5** oz

13 oz + 10 oz = 1 lb 7 oz
/\
3 7
12 oz + 12 oz = 24 oz
= 1 lb 8 oz

Practice

Task 9, p. 43

Note: Tasks 9(d) and 9(f) are more appropriate after the next lesson, so you may want to save them for later.

9. (a) **7** oz
   (b) 4 lb **12** oz
   (c) 1 lb **8** oz
   (d) 7 lb **1** oz
   (e) **9** lb **13** oz
   (f) **3** lb **14** oz

Workbook

Exercise 5, p. 40-41 (answers p. 42)

Extension

Ask your student to find the number of pounds in 48 oz. Remind him that we are going from a smaller unit of measurement to a larger one. The number of pounds will be smaller than the number of ounces. Since there are 16 ounces in a pound, we can make groups of 16 ounces and subtract from 48 ounces repeatedly to find the number of groups.

48 oz = ___________ lb

48 – 16 – 16 –16 = 0

48 oz = 3 lb

Then, have your student find the number of pounds in 39 ounces. We can either count down by 16, or count up by 16. This time there will be left-over ounces, so the answer has to be in pounds and ounces.

39 oz = ___________ lb

39 – 16 – 16 = 7

Or

16 , 32, 48, too much; 39 – 32 = 7

39 oz = 2 lb 7 oz

(2) Add and subtract in pounds and ounces

Discussion

Task 10, p. 44

> 10. (a) 5 lb 7 oz
>   (b) 1 lb 11 oz

To add or subtract pounds and ounces, we first add or subtract the pounds, and then the ounces. Remind your student that when we added kilograms and grams, we could convert the entire amounts to grams, add the grams, and then convert back to kilograms and grams. Ask her whether we could also do the same here. We can, but it is more difficult. With the metric system, it is easy to convert, because we are only multiplying by 100 or 1000. With this customary U.S. system, we have to multiply by 16 to convert to pounds and ounces, then divide by 16 to get a quotient with remainders to convert back. If we add the pounds first, then we just have to convert with 1 pound.

10(a): Discuss strategies for adding the ounces in the second step.

1) We can take 2 oz off the 9 oz to make a pound with the 14 oz.

2) We can take 7 oz from the 14 oz to make a pound with the 9 oz.

3) We can add 9 oz and 14 oz to get 23 oz and then convert.

> 1) 4 lb 9 oz + 14 oz = 5 lb 7 oz
>   (9 oz split into 7 oz and 2 oz)
>
> 2) 4 lb 9 oz + 14 oz = 5 lb 7 oz
>   (14 oz split into 7 oz and 7 oz)
>
> 3) 4 lb 9 oz + 14 oz = 4 lb 23 oz
>   = 5 lb 7 oz

10(b): Discuss strategies for subtracting the ounces in the second step.

1) We can subtract 14 oz from one of the pounds, and add back in the 9 ounces.

2) We can convert 1 lb 9 oz to 25 oz and subtract 14 oz from that.

> 1) 2 lb 9 oz – 14 oz = 1 lb 11 oz
>   (2 lb split into 1 lb and 1 lb)
>   1 lb – 14 oz = 2 oz
>   + 9 oz = 11 oz
>
> 2) 2 lb 9 oz – 14 oz = 1 lb 25 oz – 14 oz
>   = 2 lb 11 oz

Practice

Tasks 11-14, p. 45

Workbook

Exercise 6, pp. 42-43 (answers p. 42)

Reinforcement

Extra Practice, Unit 7, Exercise 3, pp. 133-134

Test

Tests, Unit 7, 3A and 3B, pp. 57-63

> 11. (a) 3 lb 7 oz + 2 lb 10 oz = 6 lb 1 oz
>   The total weight is **6 lb 1 oz.**
>   (b) 3 lb 7 – 2 lb 10 oz = 13 oz
>   The difference in weight is **13 oz.**
>
> 12. 21 lb – 12 lb 9 oz = 8 lb 7 oz
>   The smaller watermelon weighs **8 lb 7 oz.**
>
> 13. Weight of avocado: 3 oz + 4 oz = 7 oz
>   Weight of squash = 7 oz x 2 = **14 oz**
>
> 14. Weight of bananas: 21 lb ÷ 7 = 3 lb
>   Total weight: 21 lb + 3 lb = **24 lb**

Review 7

Review

Review 7, pp. 46-47

Workbook

Review 7 pp. 44-48 (answers p. 43)

Enrichment

Mental Math 7

Test

Tests, Units 1-7, Cumulative A and B, pp. 65-76

1. (a) 541 (b) 4100 (c) 3147
2. (a) 605 (b) 1724 (c) 7004
3. (a) 371 (b) 780 (c) 1628
4. (a) 29 (b) 13 (c) 54
5. 1 carton: 24 boxes
   8 cartons: 24 x 8 = 192 boxes
   There are **192** boxes in 8 cartons.
6. 5 people: $450
   1 person: $450 ÷ 5 = $90
   Each person received $**90**.
7. Group cupcakes by 7.
   140 ÷ 7 = 20
   She gave cupcakes to **20** friends.
8. 1 box: 46 buttons
   8 boxes: 46 x 8 = 368 buttons
   368 – 200 = 168 buttons
   There are **168** green buttons.
9. Number of unbroken bulbs: 150 – 6 = 144
   Group by 4's: 144 ÷ 4 = 36
   There were **36** boxes.
10. 36 + 54 + 54 = 144
    He bought **144** oranges.
11. (a) 500 cm (b) 408 cm
    (c) 2560 m (d) 3005 m
    (e) 1030 g (f) 2080 g
    (g) 50 oz (h) 1 lb 2 oz
12. (a) 2 m 8 cm
    (b) 1 km 850 m
    (c) 3 kg 95 g
13. (a) 2 m 28 cm (b) 5 m 40 cm
    (c) 65 cm (d) 2 m 20 cm
    (a) 6 km 210 m (b) 10 km 200 m
    (c) 8 km 640 m (d) 4 km 550 m
    (a) 5 kg 45 g (b) 8 kg 110 g
    (c) 1 kg 180 g (d) 1 kg 795 g
    (m) 8 lb 3 oz (n) 9 lb 6 oz
    (o) 3 lb 14 oz (p) 1 lb 2 oz
14. (a) 330 g
    (b) 330 g – 90 g = **240 g**
    (c) 240 g ÷ 2 = **120 g**

Workbook

Exercise 1. pp. 28-29

1. (a) g
   (b) g
   (c) kg
   (d) kg
   (e) g
   (f) kg
   (g) g; kg

2. (a) 2 kg 500 g (b) 1 kg 200 g

3. (a) 1 kg 400 g (b) 2 kg 700 g
   (c) 1 kg 700 g (d) 3 kg 700 g

Exercise 2, pp. 30-33

1.

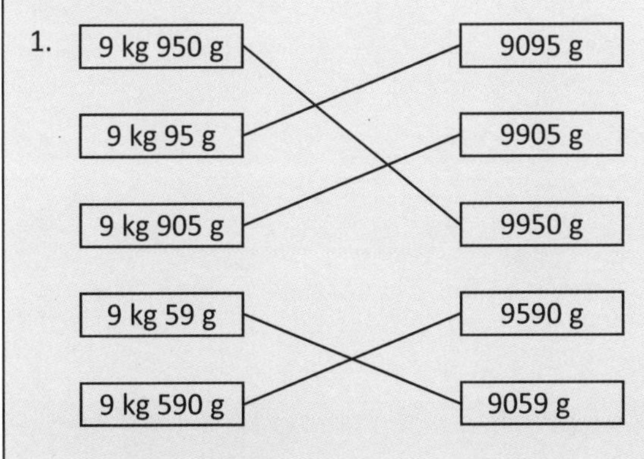

2. 1 kg 10 g → 1010 g
   1 kg 100 g → 1100 g
   1 kg 250 g → 1250 g
   1 kg 25 g → 1025 g
   2 kg 25 g → 2025 g
   2 kg 50 g → 2050 g
   3 kg 80 g → 3080 g
   3 kg 8 g → 3008 g

3. (a) 1800
   (b) 6020
   (c) 2300
   (d) 9002
   (e) 4083
   (f) 8015

4. (a) 1 kg 280 g
   (b) 4 kg 69 g
   (c) 2 kg 506 g
   (d) 5 kg 108 g
   (e) 3 kg 9 g
   (f) 6 kg 4 g

5. (a) 3005 g $<$ 3050 g (b) 2020 g = 2020 g
   (c) 4008 g $<$ 4010 g (d) 1086 g = 1086 g

5. (a) hen; duck (1550 g $>$ 1050 g)
   (b) A, B (2090 g $<$ 2780 g)

6. A: 1067 g B: 764 g C: 2670 g D: 2700 g
   (a) D
   (b) B
   (c) B
   (d) D

Exercise 3, pp. 34-36

1. (a) 1 kg (b) 200 g
   (c) 550 g (d) 330 g
   (e) 250 g
   (f) 610 g
   (g) 850 g
   (h) 780 g

2. (a) 1 kg 850 g
   (b) 3 kg 250 g
   (c) 4 kg 280 g

3. (a) 3 kg 765 g
   (b) 6 kg 250 g
   (c) 6 kg 55 g
   (d) 8 kg 9 g

4. (a) 4 kg 90 g
   (b) 4 kg 545 g
   (c) 6 kg 635 g

5. (a) 1 kg 156 g
   (b) 2 kg 742 g
   (c) 850 g
   (d) 6 kg 736 g

Workbook

Exercise 4, pp. 37-39

1. (a) **280 g**
   (b) 280 g – 100 g = **180 g**

2. (a) **370 g**
   (b) 370 g – 150 g = **220 g**

3. (a) **330 g**
   (b) 330 g – 130 g = **200 g**
   (c) 200 g ÷ 2 = **100 g**

4. (a) **330 g**
   (b) 60 g x 2 = **120 g**
   330 g – 120 g = **210 g**
   (c) 210 g ÷ 5 = **42 g**

5. (a) 2 kg 50 g + 600 g = 2 kg 650 g
   The watermelon weighs **2 kg 650 g**.
   (b) 2 kg 50 g + 2 kg 650 g = 4 kg 700 g
   The total weight is **4 kg 700 g**.

6. 2 kg 400 g – 1 kg 950 g = 450 g
   The bag of salt weighs **450 g.**

Exercise 5, pp. 40-41

1. (a) 3 lb 8 oz (b) 6 lb 10 oz

2. (a) 32 oz (b) 58 oz (c) 137 oz

3. (a) 1 lb 2 oz (b) 1 lb 6 oz (c) 2 lb 0 oz

4. (a) 22 oz **>** 21 oz (b) 155 oz **<** 157 oz
   (c) 16 oz **=** 16 oz (d) 20 oz **>** 18 oz

5. (a) 16 oz
   (b) 13 oz
   (c) 6 oz
   (d) 7 oz
   (e) 4 lb 2 oz
   (f) 3 lb 14 oz
   (g) 1 lb
   (h) 1 lb 3 oz
   (i) 2 lb 3 oz
   (j) 1 lb 2 oz
   (k) 2 lb 7 oz
   (l) 8 oz

6. 8 lb – 4 oz = 7 lb 12 oz
   The difference in weight is **7 lb 12 oz**.

Exercise 6, pp. 42-43

1. (a) 13 lb 1 oz
   (b) 12 lb 0 oz
   (c) 7 lb 1 oz

2. (a) 2 lb 6 oz
   (b) 1 lb 14 oz
   (c) 0 lb 15 oz

3. (a) 3 lb
   (b) 3 lb: $9
   1 lb: $9 ÷ 3 = $3
   1 pound costs **$3**.

4. 7 lb 10 oz + 9 lb 7 oz = 17 lb 1 oz
   The total weight is **17 lb 1 oz**.

5. 20 lb 7 oz – 13 lb 9 oz = 6 lb 14 oz
   The smaller watermelon weighs **6 lb 14 oz**.

6. 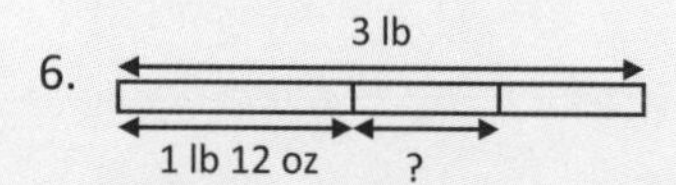

   2 cheese: 3 lb – 1 lb 12 oz = 1 lb 4 oz = 20 oz
   1 cheese: 20 oz ÷ 2 = 10 oz
   One package of cheese weighs **10 oz**.

7. (a) 3 lb 14 oz – 10 oz = 3 lb 4 oz
   Carson's berries **weigh 3 lb 4 oz**.
   (b) 4 lb 6 oz – 10 oz = 3 lb 12 oz
   Maureen's berries weigh **3 lb 12 oz**.
   (c) 3 lb 4 oz + 3 lb 12 oz = 7 lb
   Together, they picked **7 lb** of berries.

Workbook

Review 7, p. 44-48

1. (a) 60
   (b) 300
   (c) 70
   (d) 500
2. (a) x
   (b) –
   (c) ÷
   (d) +
   (e) –
3. $45 x 9 = $405
   She earned **$405.**
4. 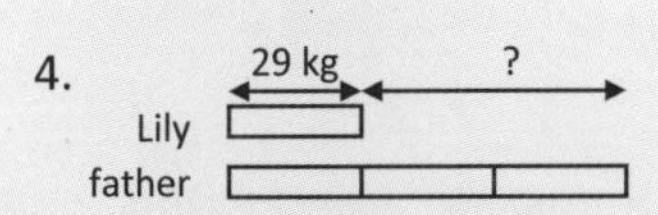

   1 unit = 29 kg
   2 units = 29 kg x 2 = 58 kg
   Lily's father is **58 kg** heavier than Lily.
5. 2500 – 1164 – 940 = 396
   396 tiles were left over.
6. 

   Number of women: 2000 – 1340 = 660
   Difference: 1340 – 660 = 680
   There are **680** more men than women.
7. 3 km 600 m – 2 km 800 m = 800 m
   Jim jogged **800 m** farther.
8. Cost for 4: $100 – $48 = $52
   Cost for 1: $52 ÷ 4 = 13
   Each person paid **$13.**
9. Total cookies: 8 x 12 = 96
   96 – 28 = 68
   68 cookies were eaten.
10. 5 kb – 200 g = 4 kg 800 g
    The sand weighs **4 kg 800 g**.
11. 4 ft 1 in. – 5 in. = 3 ft 8 in.
    His brother is **3 ft 8 in.** tall.
12. 3 units = 240
    1 unit = 240 ÷ 3 = 80
    2 units = 80 x 2 = 160
    The numbers are **80** and **160.**
13. (a) David
    (b) Andy
14. (a) 1 head, 1 tail: ~~||||~~ ~~||||~~ ~~||||~~ ~~||||~~ ~~||||~~ |||
    (b) Check graph, bar should go to 28.

Unit 8 – Capacity

Chapter 1 – Liters and Milliliters

Objectives

- Measure capacity in liters and milliliters.
- Convert between liters and milliliters.
- Add and subtract liters and milliliters in compound units.

Vocabulary

- Capacity
- Liters
- Milliliters

Notes

In *Primary Mathematics* 2B, the liter is introduced as a standard unit of measurement. In this chapter, the milliliter will be introduced and students will measure in liters and milliliters, convert measurements between liters and milliliters, and add and subtract measurements in liters and milliliters. 1 liter = 1000 milliliters.

Give your student adequate practical experience in measuring capacity so that she can make reasonable estimates of the capacity of containers.

For your information, the SI unit for volume is a cubic meter, that is, the amount of space equivalent to a cube that is a meter long on each side. Most countries, however, measure volume in liters. One liter is equal to a cubic decimeter, or 1,000 cubic centimeters, and there are 1,000 liters in a cubic meter. Some countries measure capacity in deciliters. There are 10 deciliters in a liter.

A liter is a little more than a quart (1.057 U.S. liquid quarts). A measuring teaspoon has a capacity of about 5 milliliters, and a 1 cup measuring cup has a capacity of about 250 milliliters.

In the textbook the liter is abbreviated with a cursive ℓ. It is often abbreviated with a capital L in other texts. Your student can use a capital L, rather than trying to write the cursive ℓ. In this textbook, the abbreviation for milliliters is ml. In some texts that use a cursive ℓ it is abbreviated as mℓ, and in texts that use a capital L for liters it is abbreviated as mL.

The capacity of a container is how much water it can hold. In the last chapter of *Primary Mathematics* 3B, students will be introduced to the term volume, which is the amount of space an object occupies. In this chapter, students will be measuring different volumes of liquid, but you do not have to introduce the term volume yet.

Material

- Quart/liter measuring cup
- Cup or pint measuring cup with marks for milliliters
- Measuring teaspoon
- Optional: 1000 ml, 500 ml, 100 ml beakers
- Paper cups or regular drinking cups
- Various empty containers for liquids

(1) Measure in liters and milliliters

Activity

Show your student a measuring cup and discuss the markings for milliliters. Tell him that milliliters are used to measure smaller amounts than a liter. Have him use the measuring cup to tell you how many milliliters are in a liter. Guide him in measuring specific amounts of water by filling the measuring cup up to the required amount. Use amounts that are marked on the measuring cup.

Discussion

Concept page 48

(a) 750 ml
(b) 2 ℓ 300 ml

Point out the abbreviation for liters and milliliters. Tell your student that he can use a capital L for liters. If needed, discuss the scales of the markings on each type of beaker.

Task 1, p. 49

1. 2 liters

Make sure your student understands that the capacity is the amount a container can hold when full. Usually the capacity of liquid containers is measured to some fill mark, not necessarily to the very top. So the capacity of the liter beakers above is 1 liter, even though a bit more could be added.

Tasks 2-8, pp. 49-52

4. (c) 1000 ml

5. 5 ml

Do as many of these activities as you feel would be useful, and adapt according to the materials you have. The purpose is to give your student an understanding of how much a liter and a milliliter is and an ability to roughly estimate the capacity of various containers. You student does not need to fill the paper cups, for example, until they can hold absolutely no more water, just to almost the top. We don't need to know the capacity of a bucket to the closest milliliters, just about how many liters it can hold.

Unless you have a graduated cylinder or a small beaker, your student cannot measure 10 milliliters in task 5, which would be about 2 teaspoons. He may be able to measure 50 milliliters with a measuring cup (about 10 teaspoons) and divide 50 milliliters by 10 to find that a teaspoon is about 5 milliliters.

Practice

Task 9, p. 52

9. (a) 350 ml
(b) 800 ml
(c) 1 ℓ 200 ml

Workbook

Exercise 1, pp. 49-50 (answers p. 55)

(2) Convert between liters and milliliters

Discussion

Tasks 10-11, p. 53

Tell your student that the conversion factor for liters to milliliters is 1000 ml/L, that is, 1000 milliliters in 1 liter. Each thousand milliliters is a liter.

10. 1100 ml
    1 ℓ 100 ml

11. 1 ℓ 500 ml

Practice

Task 12 p. 53

12. (a) 1 ℓ 200 ml (b) 2 ℓ 500 ml (c) 2 ℓ 50 ml
    (d) 1 ℓ 5 ml (e) 3 ℓ 400 ml (f) 3 ℓ 105 ml

Discussion

Task 13, p. 53

We can convert liters to milliliters simply by multiplying the liters by 1000.

13. (a) 2000 ml
    (b) 2350 ml

Practice

Tasks 14-16, p. 54

14. (a) 1800 ml (b) 1080 ml (c) 1008 ml
    (d) 3025 ml (e) 2005 ml (f) 3500 ml

Discussion

Tasks 15-17

15. 1 ℓ 250 ml

16. 350 ml

17. A; 260 ml

15: See if your student can come up with a quick way to do this mentally. Tell her that it is useful to memorize 250 x 4 = 1000. She should be familiar with 25 x 4 = 100 from knowing that four quarters makes a dollar. So 25 tens x 4 is 100 tens, or 1000. Other useful facts to memorize are 20 x 5 = 100 and 200 x 5 = 1000.

16: Your student can do this problem easily by simply seeing from the scale on the last beaker that 350 liters need to be added. You may want to write the equation 1 ℓ 650 + ____ = 2 ℓ to remind him that we are making the next 1000.

1 ℓ 650 ml + ____ = 2 ℓ

1000 ml – 650 ml = 350 ml

17: Write the subtraction expression and see if your student can solve it mentally by subtracting 780 from 1000 and adding 40.

```
  2 ℓ 40 ml – 1 ℓ 780 ml
  /\
1 ℓ  1 ℓ – 780 ml = 220 ml
                  + 40 ml = 260 ml
```

Workbook

Exercise 2, pp. 51-53 (answers p. 55)

(3) Add and subtract liters and milliliters

Discussion

Task 18, p. 55

To add or subtract liters and milliliters, we first add or subtract the liters, and then the milliliters.

18. (a) 5 ℓ 150 ml
    (b) 1 ℓ 550 ml

10(a): Discuss strategies for adding the milliliters in the second step.

1) We need 200 ml more to make 1 liter with the 800 ml. That leaves 150 ml.
2) We can add and then convert.
3) Instead of adding liters and then milliliters, we can convert the whole measurement to ml, add, and convert back.

1) 4 ℓ 800 ml + 350 ml = 5 ℓ 150 ml
   / \
   200 ml 150 ml

2) 4 ℓ 800 ml + 350 ml = 4 ℓ 1150 ml
   = 5 ℓ 150 ml

3)
```
  1 8 0 0
+ 3 3 5 0
---------
  5 1 5 0  → 5 ℓ 150 ml
```

10(b): Discuss strategies for subtracting the ounces in the second step.

1) We can subtract 800 ml from a liter and add back in the 350 ml.
2) We can convert 1 ℓ 350 ml to 1350 ml and subtract 800 ml.
3) Instead of subtracting liters and then milliliters, we can convert the whole measurement to milliliters, subtract, and convert back.

1) 2 ℓ 350 ml – 800 ml = 1 ℓ 550 ml
   / \
   1 ℓ 1 ℓ – 800 ml = 200 ml
   + 350 ml = 550 ml

2) 2 ℓ 350 ml – 800 ml = 1 ℓ 1350 ml – 800 ml
   = 1 ℓ 550 ml

3)
```
  3 3 5 0
– 1 8 0 0
---------
  1 5 5 0  → 1 ℓ 550 ml
```

Practice

Tasks 19-20, p. 56

Workbook

Exercise 3, pp. 54-58 (answers p. 55)

19. (a) 2 ℓ (b) 4 ℓ
    (c) 4 ℓ 50 ml (d) 9 ℓ 140 ml
    (e) 1 ℓ 20 ml (f) 2 ℓ 150 ml
    (g) 2 ℓ 720 ml (h) 3 ℓ 925 ml

20. (a) Container A
    (b) Container B
    (c) 8 ℓ 30 ml

(4) Practice

Practice

Practice A, p. 57

Reinforcement

Extra Practice, Unit 8, Exercise 1, pp. 139-142

Test

Tests, Unit 8, 1A and 1B, pp. 77-83

1. (a) 3000 ml (b) 1200 ml (c) 2055 ml
   (d) 2650 ml (e) 3065 ml (f) 4005 ml

2. (a) 5 ℓ (b) 1 ℓ 600 ml (c) 2 ℓ 250 ml
   (d) 3 ℓ 205 ml (e) 2 ℓ 74 ml (f) 1 ℓ 9 ml

3. (a) 1000 > 980
   (b) 2050 = 2050
   (c) 4008 < 4800

4. (a) 2 ℓ 650 ml + 5 ℓ 300 ml = 7 ℓ 950 ml
   The total capacity is **7 ℓ 950 ml.**
   (b) 5 ℓ 300 ml – 2 ℓ 650 ml = 2 ℓ 650 ml
   B can hold **2 ℓ 650 ml** more than A.

5. 2 ℓ 800 ml + 1 ℓ 600 ml = 4 ℓ 400 ml
   Y holds **4 ℓ 400 ml** of water.

6. 2 ℓ x 9 = 18 ℓ
   The capacity is **18 ℓ.**

Enrichment

See if your student can solve the following problem. Encourage him to draw a diagram:

⇒ There are 3 bottles with the following capacities and no other marks:

4 liters
3 liter
1 liter

The 4 liter bottle is full.

How can you use only the 3 bottles to divide up the water so that the 4-liter and the 3-liter bottle both have 2 liters?

Possible answer: Pour from the 4 liter bottle into the 3 liter-bottle, leaving 1 liter in the 4-liter bottle. Then pour from the 3-liter bottle into the 1-liter bottle, leaving 2 liters in the 3-liter bottle. Then pour from the 1-liter bottle back into the 4-liter bottle, which will now have 2 liters.

Chapter 2 – Gallons, Quarts, Pints and Cups

Objectives

- Estimate and measure capacity in compound units, using gallons, quarts, pints, and cups.
- Convert between gallons, half-gallons, quarts, pints, and cups.
- Add and subtract capacity measurements in compound units.

Vocabulary

- Gallon
- Half-gallon
- Quart
- Pint
- Cup

Notes

The customary units for capacity used in the U.S. were introduced in *Primary Mathematics* 2A. In this chapter, students will measure capacity in compound units, convert from a compound unit to a single unit or vice versa, and add and subtract in compound units.

Give your student practical experience in measuring capacity. Cooking and baking are a familiar use for these units of capacity, so you could have your student measure quantities for recipes.

The strategies already learned in converting between measurements and adding and subtracting compound measurements also apply here, except that the conversion factors are 2 or 4.

You may want to extend to discussions to include tablespoons and teaspoons used for measuring (not the ones used for eating), and fluid ounces. There are 3 teaspoons in a tablespoon and 4 tablespoons in a fourth of a cup. Measuring cups may also be marked in fluid ounces. A fluid ounce is another customary unit of measuring volume. There are 8 fluid ounces in a cup.

1 gallon	=	2 half-gallons
1 half-gallon	=	2 quarts
1 quart	=	2 pints
1 pint	=	2 cups
1 gallon	=	4 quarts
	=	8 pints
	=	16 cups
1 quart	=	4 cups

1 cup	=	16 tablespoons
1 tablespoon	=	3 teaspoons
1 cup	=	8 fluid ounces
1 liter	=	1.057 quarts
1 liter	=	33.8140 fluid oz
1 cup	=	about 240 ml

Material

- Glass or plastic measuring cups with pints and cups marked
- Gallon and half-gallon container
- Various small containers
- Multilink cubes
- 4 sets of number cards 1-10.
- 5 cards each with **gal**, **qt**, **pt**, and **c** written on them – 20 unit cards total.
- Mental Math 8 (Appendix)

(1) Convert between gallons, quarts, pints, and cups

Discussion

Concept page 58
Task 1, p. 58

1 gal	=	2 half-gal
1 half-gal	=	2 qt
1 qt	=	2 pt
1 pt	=	2 c

Tell your student that he should try to remember the order gallon → half-gallon → quart → pint → cup and that the conversion factor going from each to the next is 2. This will make all other conversions, such as gallon to pint, easy to figure out. Point out the abbreviations for gallon (gal), quart (qt), pint (pt), and cup (c). Tell him that a half-gallon is not really a unit of measurement, and does not have its own abbreviation, but putting it in the sequence will help him remember the number of quarts in a gallon.

Activity

Show your student a gallon container, a half-gallon containers, and any other containers you might have that have capacities of a quart, a pint, and a cup, such as various dairy products, and measuring cups.

Discuss the number of cups in a pint, pints in a quart, quarts in a half-gallon, half-gallons in a gallon. Ask her to fill the measuring cups with water to various amounts, such as 1 pint and 1 cup or 3 cups. Have her estimate and measure the capacity of a few containers. You can also show her teaspoons and tablespoons used for measuring when cooking (not the ones used for eating) and tell her that there are 3 measuring teaspoons in a measuring tablespoon, and 4 measuring tablespoons in a quarter cup. Remind her that a quart is almost the same as a liter, and that a cup is close to 250 milliliters.

You may want to point out the markings for fluid ounces on measuring cups. Tell your student that a fluid ounce is another measurement for capacity. There are 8 fluid ounces in a cup. Various recipes, especially on containers with drink mixes, use fluid ounces. For example, a hot cocoa might say to put 2 tablespoons of hot cocoa in a mug and then add 6-8 fluid ounces of hot water. This would fill up a regular sized mug.

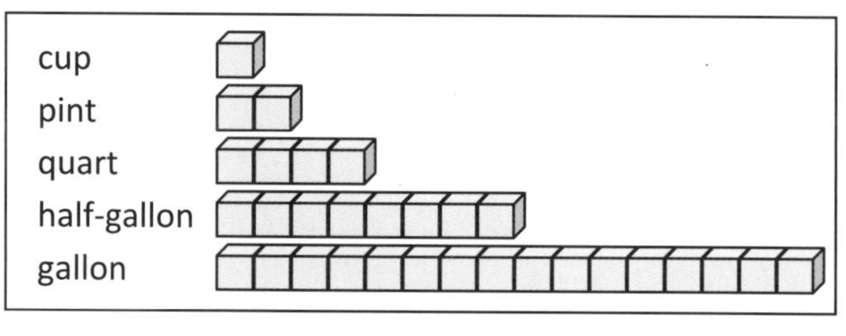

Optional: Use multilink cubes. Tell your student that 1 cube represents 1 cup and have him represent the other measurement units with the cubes. He can set them on paper and label each in order to use them to help him with conversions. You can also use Legos™. A piece with 1 knob represents a cup, a piece with 2 knobs represents a pint, a piece with 4 knobs represents a quart, and a piece with 16 knobs represents a gallon.

Discussion

Task 2, p. 59

1 qt	=	4 c
1 gal	=	4 qt
	=	8 pt
	=	16 c

Ask your student to also find the capacity of a gallon in quarts, pints, and cups.

Tasks 3-7, pp. 59-60

Have your student solve these, and then discuss methods of solution.

3: To convert 5 quarts to gallons and quarts, we make a group of 4 quarts, which is a gallon, and have 1 quart left over.

4: To convert 78 quarts to gallons and quarts, we also make groups of 4. So we divide by the conversion factor, 4 qt/gal. The quotient is the number of gallons, and the remainder is the number of quarts. Remind your student that since a gallon contains a larger quantity than a quart, the equivalent number of gallons will be a smaller number; that is, every 4 quarts is 1 gallon.

5,7: Every 2 pints is a quart, so we are making groups of 2. We can divide by 2 to convert pints to quarts. The quotient is the number of quarts, and the remainder is the left over pints.

6: Since there are 2 cups in a pint, we can again divide by 2 to convert to pints and cups.

3. 5 qt
   1 gal 1 qt

4. 78 ÷ 4 = 19 R 2
   78 qt = **19 gal 2 qt**

$$\begin{array}{r} 19 \\ 4\overline{)78} \\ \underline{4} \\ 38 \\ \underline{36} \\ 2 \end{array}$$

5. 15 ÷ 2 = 7 R 1
   15 pt = **7 qt 1 pt**

6. 21 ÷ 2 = 10 R 1
   21 c = **10 pt 1 pt**

7. 2 qt 1 pt

Activity

Have your student do the following problems.

⇒ 4 gal 3 qt = ____ qt

⇒ 9 qt 1 p = ____ c

4 gal 3 qt= ____ qt

4 gal x 4 qt/gal = 16 qt
+ 3 qt = 19 qt

4 gal 3 qt= 19 qt

9 qt 1 p = ____ c

9 qt x 4 c/qt = 36 c
1 pt x 2 c/pt = 2 c

9 qt 1 p = 38 c

Workbook

Exercise 4, pp. 59-60 (answers p. 56)

Game

Material: 4 sets of number cards 1-10. 5 cards each with **gal**, **qt**, **pt**, and **c** written on them – 20 unit cards total. (Make 20 cards for 2 players. If there are 3 players make 6 cards of each; if there are 4 players make 7 cards of each, etc.)

Procedure: Shuffle the number cards and place them in the middle face down. Shuffle the unit cards and place them in the middle face down. Each player turns over a number card and two unit cards from the middle. The player must convert the number from the largest volume unit to the smallest. For example, the player turns over a 4, a **c** and a **qt**. She must convert 4 qt to cups. The result is 16. The player with the greatest result gets a point. If two players have the same result then they draw again to see who gets the greatest result. The play continues until all number cards have been turned over. When all the unit cards have been turned over they are shuffled and placed upside down again. The player with the most cards at the end wins.

(2) Add and subtract gallons, quarts, pints, and cups

Discussion

Tasks 8-9, textbook p. 61

The procedure used here is similar to what your student has already learned; add one unit first, then the other, converting if needed. If needed, you can discuss various strategies with task 9.

8. (a) 6 pt
   3 qt
   (b) 1 pt 1 c; 1 pt

9. (a) 30 gal 1 qt
   (b) 0 gal 3 qt

9(a): Add the gallons first.

1) We can add quarts by "making 1 gallon", which leaves 1 qt.
2) Or we can add quarts and then convert.

15 gal 2 qt + 14 gal = 29 gal 2 qt

1) 29 gal 2 qt + 3 qt = 30 gal 1 qt
   (3 qt split into 2 qt and 1 qt)

2) 29 gal 2 qt + 3 qt = 29 gal 5 qt
   = 29 gal 1 gal 1 qt
   = 30 gal 1 qt

9(b): We subtract the gallons first.

1) We can subtract the 3 quarts from the gallon, leaving 1 qt, and add back in the 2 qt.
2) We can convert 1 gal 2 qt to 6 qt and then subtract 3 qt.

15 gal 2 qt – 14 gal = 1 gal 2 qt

1) 1 gal 2 qt – 3 qt = 3 qt
   1 gal – 3 qt = 1 qt
   + 2 qt = 3 qt

2) 1 gal 2 qt – 3 qt = 6 qt – 3 qt
   = 3 qt

Practice

Task 10, p. 61

10. (a) 19 pt
    (b) 22 gal 2 qt
    (c) 1 gal 2 qt
    (d) 20 qt 1 c

Workbook

Exercise 5, p.61 (answers p. 56)

(3) Practice

Practice

Practice B, p. 62

Reinforcement

Extra Practice, Unit 8, Exercise 2, pp. 143-146

Mental Math 8

Test

Tests, Unit 8, 2A and 2B, pp. 85-90

Enrichment

Write the following problem and have your student work on it. Then have her explain her solution. A suggested method is given.

⇒ Two tanks have the same amount of water. If 1 quart and 1 pint of water is poured from Tank A into Tank B, Tank B will be full and Tank A will be half-full. What is the total capacity of both tanks, in gallons?

Use bar diagrams to draw the situation after the water has been transferred. From the diagram, we can see that if they started out with the same amount, then half of the difference, that is, half of the unit, must have been poured from A into B. Divide all the units into half. B has 4 units and A has 2 units; they both started with 3 units, and 1 unit was poured from A to B. If B is now full, and A is half-full, then both tanks must have a capacity of 4 units, and altogether they have a capacity of 8 units.

1. (a) 16 c (b) 31 c
2. (a) 14 pt (b) 23 pt
3. (a) 40 qt (b) 93 qt
4. (a) 21 qt > 12 qt
   (b) 25 qt > 23 qt
   (c) 15 c = 15 c
5. (a) 11 qt 0 pt
   (b) 4 gal 3 qt
   (c) 72 pt 1 c
6. 13 gal – 7 gal 1 qt = 5 gal 3 qt
   B holds **5 gal 3 qt**.
7. 2 qt – 1 pt = 4 pt – 1 pt = 3 pt
   3 pints more water is needed.
8. 2 c x 7 = 14 c = 7 pt
   She drinks **7 pt** in a week.

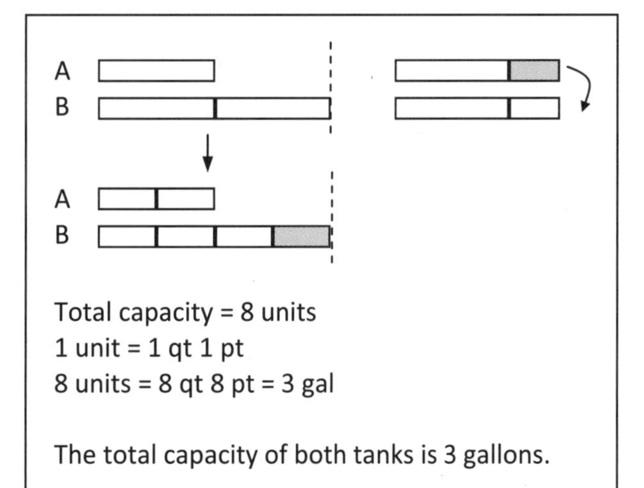

Total capacity = 8 units
1 unit = 1 qt 1 pt
8 units = 8 qt 8 pt = 3 gal

The total capacity of both tanks is 3 gallons.

Review 8

Review

Review 8, pp. 63-65

Your student should be able to solve the first 11 problems mentally, and possibly be able to use mental computation with the rest of the problems.

12: You can supply the answer for this one if you wish, rather than requiring your student to memorize the conversion factor for feet to miles.

15-19: If your student has used earlier levels of *Primary Mathematics*, he will probably not have to draw bar models for these problems.

Workbook

Review 8 pp. 62-66 (Answers p. 56)

Test

Tests, Units 1-8, Cumulative A and B, pp. 91-104

1. (a) 150 (b) 300 (c) 180
2. (a) 20 (b) 20 (c) 200
3. (a) 540 (b) 2800 (c) 2400
4. (a) 30 (b) 40 (c) 80
5. $50 x 8 = $400
   She saves **$400** in 8 months.
6. 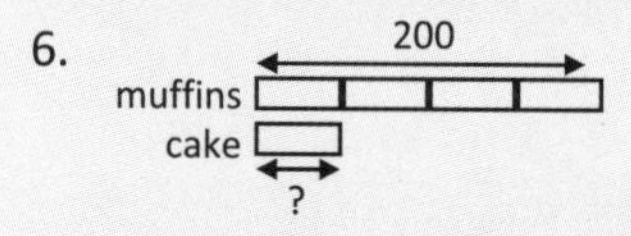

   4 units = 200
   1 unit = 200 ÷ 4 = 50
   It sold **50** chocolate cakes.
7. 9 x 40 = 360
   He bought **360** pears.
8. 6 x 200 = 1200
   There are **1200** coins in 200 sets.
9. (a) 250 ÷ 5 = 50
   There were **50** bags.
   (b) $2 x 50 = $100
   He received **$100**.
10. (a) 98 + 62 = 160
    He bought **160** pens.
    (b) 160 ÷ 8 = 20
    There were **20** pens in each box.
11. (a) 909 cm (b) 4 m 5 cm
    (c) 4009 m (d) 4 km 1 m
12. 5280 feet
13. 114 in. (b) 70 in. (c) 91 in.
14. (a) 6 lb 9 oz
    (b) 14 ft 1 in.
    (c) 19 qt
    (d) 5 lb 6 oz
    (e) 2 ft 10 in.
    (f) 6 gal 3 qt
15. 24 ÷ 3 = 8
    8 buckets are needed.
16. 8 ℓ – 4 ℓ 650 ml = 3 ℓ 350 ml
    3 ℓ 350 ml more water is needed.
17. Total paint: 3 ℓ x 6 = 18 ℓ
    18 ℓ – 2 ℓ 400 ml = 15 ℓ 600 ml
    He used **15 ℓ 600 ml** of paint.
18. 1 lb 5 oz + 14 oz = 2 lb 3 oz
    The box of peaches weighs **2 lb 3 oz.**
    1 lb 5 oz + 2 lb 3 oz = 3 lb 8 oz
    Together they weigh **3 lb 8 oz**.
19. 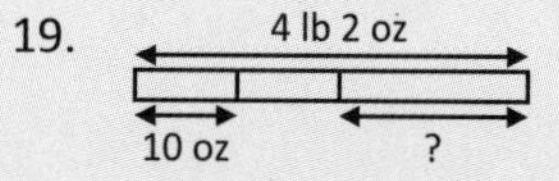

    Weight of sugar: 20 oz = 1 lb 4 oz
    4 lb 2 oz – 1 lb 4 oz = 2 lb 14 oz
    Or: 4 lb 2 oz – 10 oz = 3 lb 8 oz
    3 lb 8 oz – 10 oz = 2 lb 14 oz
    The bag of flour weighs **2 lb 14 oz**.

Workbook

Exercise 1, pp. 49-50

1.

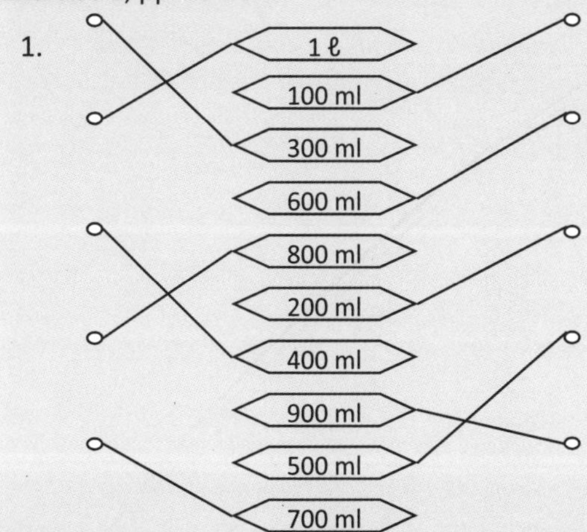

2. (a) 400 ml (b) 700 ml
   (c) 350 ml (d) 150 ml
   (e) 30 ml (f) 800 ml

Exercise 2, pp. 51-53

1.

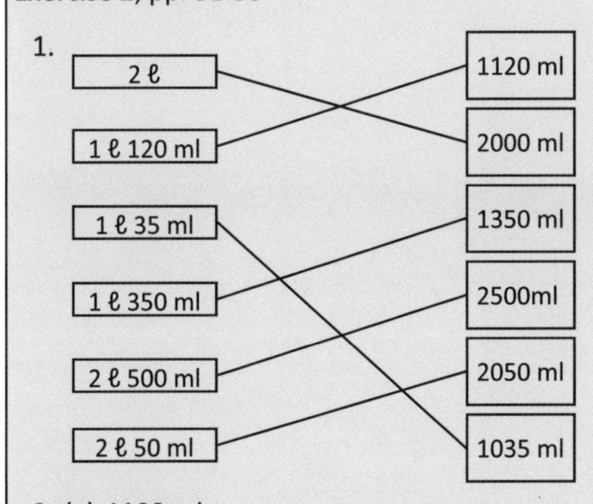

2. (a) 1100 ml
   (b) 1725 ml
   (c) 1640 ml
   (d) 2855 ml
   (e) 2025 ml
   (f) 3005 ml

3. (a) 1 ℓ 300 ml
   (b) 1 ℓ 450 ml
   (c) 2 ℓ 90 ml
   (d) 2 ℓ 105 ml
   (e) 3 ℓ 75 ml
   (f) 4 ℓ 5 ml

4. (a) 1650 ml > 1065 ml (b) 2075 ml < 2750 ml
   (c) 3030 ml = 3030 ml (d) 3090 ml < 3900 ml
   (e) 4010 ml < 4100 ml

5. 890 ml → 110 ml 725 ml → 275 ml
   495 ml → 505 ml 645 ml → 355 ml

Exercise 3, pp. 54-58

1. (a) 930 ml (b) 450 ml
   (c) 370 ml (d) 920 ml

2. (a) 140 ml
   (b) 580 ml
   (c) 250 ml
   (d) 660 ml

3. (a) 1 ℓ 750 ml
   (b) 3 ℓ 0 ml
   (c) 4 ℓ 150 ml
   (d) 5 ℓ 490 ml

4. (a) 3 ℓ 760 ml
   (b) 3 ℓ 890 ml
   (c) 5 ℓ 70 ml
   (d) 6 ℓ 45 ml
   (e) 8 ℓ 14 ml

5. (a) 3 ℓ 180 ml
   (b) 4 ℓ 40 ml
   (c) 4 ℓ 670 ml
   (d) 5 ℓ 950 ml

6. (a) 2 ℓ 180 ml
   (b) 1 ℓ 64 ml
   (c) 2 ℓ 665 ml
   (d) 0 ℓ 760 ml
   (e) 5 ℓ 721 ml

7. (a) 3 ℓ – 1 ℓ 50 ml = **1 ℓ 950 ml**
   (b) 3 ℓ + 1 ℓ 50 ml = **4 ℓ 50 ml**

8. (a) 5 ℓ – 3 ℓ = **2 ℓ**
   (b) 5 ℓ + 2 ℓ + 3 ℓ = **10 ℓ**
   (c) 10 ℓ – 8 ℓ 400 ml = **1 ℓ 600 ml**

9. 375 ml x 6 = 2250 ml = 2 ℓ 250 ml
   The total amount of milk is **2 ℓ 250 ml.**

10. Total paint = 3 ℓ x 4 = 12 ℓ
    Paint used = 12 ℓ – 2 ℓ 450 ml = **9 ℓ 550 ml**

Workbook

Exercise 4, pp. 59-60

1.

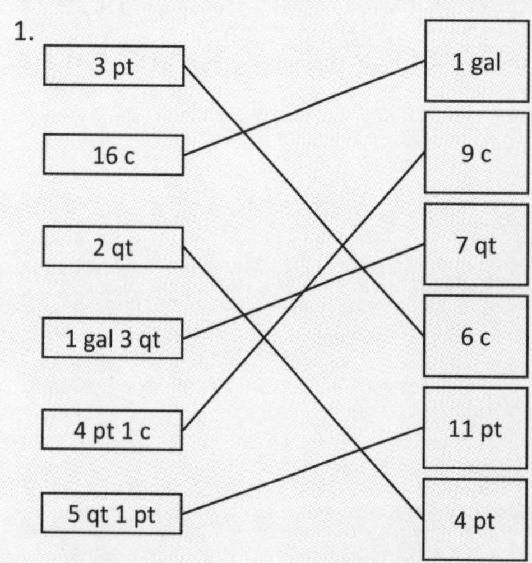

2. (a) 10 qt = 10 qt (b) 18 c > 16 c
   (c) 15 c < 16 c (d) 26 qt > 25 qt

3. 1 qt 4 pt 1 c 1 gal 2 c 20 c

4. Check marks for
   (a), (d), (e), and (g)

Exercise 5, p. 61

1. (a) 1 qt
   (b) 2 gal 2 qt
   (c) 1 gal 1 qt
   (d) 5 gal 1 qt
   (e) 1 c
   (f) 6 c

2. (a) 9 pt 0 c
   (b) 15 gal 2 qt

3. (a) 1 gal 2 qt
   (b) 6 qt 1 pt

4. (a) 15 gal 2 qt – 6 gal 3 qt = **8 gal 3 qt**
   (b) 15 gal 2 qt + 6 gal 3 qt = **22 gal 1 qt**

Review 8, pp. 62-66

1. (a) 3010 (b) 6000 (c) 4015
   (d) 2308 (e) 1968 (f) 2354

2. (a) 406 (b) 848 (c) 2304
   (d) 28 (e) 38 (f) 50 R 2

3.

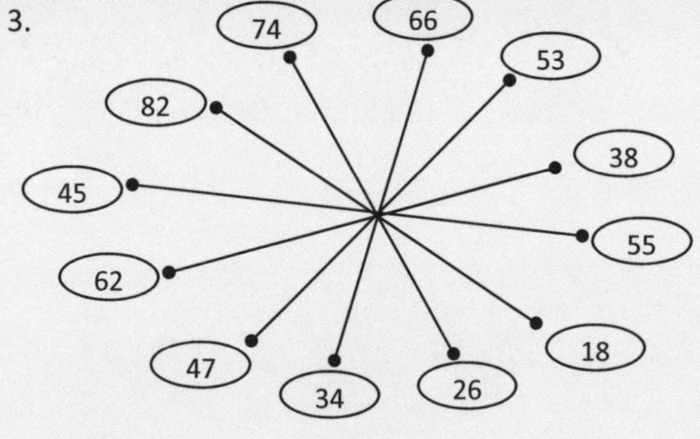

4. (a) 310 cm (b) 2 m 85 cm
   (c) 4050 g (d) 3 kg 50 g
   (e) 2005 m (f) 2 km 500 m
   (g) 3060 ml (h) 4 ℓ 5 ml

5. (a) 7 m 45 cm
   (b) 3 km 985 m
   (c) 5 kg 170 g
   (d) 2 ℓ 960 ml

6. (a) 1 ℓ 400 ml
   (b) 1 ℓ
   (c) 200 ml

7. 3 km 300 m + 1 km 500 m = 4 km 800 m
   Jessica's house is **4 km 800 m** from the school.

8. 6523 – 3806 = **2717**
   There are **2717** females.

9. 6 lb 8 oz + 12 oz = 7 lb 4 oz
   She used **7 lb 4 oz** of flour.

10.

402 – 35 = 367
367 + 402 = 769

He sold **769** mangoes altogether.

11.

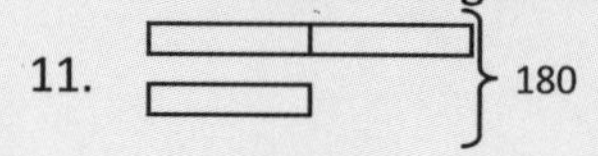

3 units = 180
1 unit = 180 ÷ 3 = 60
2 units = 60 x 2 = 120

The greater number is **120**.

12. (a) ☑ The spinner lands on a 2-digit number.
    (b) ☑ The spinner lands on a number less than 20.
    (c) ☑ The spinner lands on a number that can be divided evenly by 4.

Unit 9 – Money

Chapter 1 – Dollars and Cents

Objectives

- Review decimal notation for money.
- Count money up to $100 in coins and bills.
- Write amounts of money in words or figures.
- Convert dollars and cents to cents.
- Convert cents to dollars and cents.
- Make change for bills up to $100.

Notes

In *Primary Mathematics* 2A, students learned to count, read, and write money up to $10, to convert from dollars and cents to cents and vice-versa, and to make change for $1, $5, and $10. These concepts are reviewed here and extended to amounts of money up to $100. Students will also write amounts of money up to $10,000 in words.

The concept of decimals has not yet been taught. Decimals will be learned in *Primary Mathematics* 4B. In the meantime, the decimal point should be presented as a dot separating dollars from cents.

In writing amounts of money in words as the amount in dollars, get your student to write "zero dollars" if there are no dollars, and "and zero cents" if there are no cents. $405.00 would be written as "four hundred five dollars and zero cents." $0.45 is "zero dollars and forty-five cents."

This guide includes a lesson on making change for dollar amounts up to $100. Experience with making change by counting up from the cost amount to the amount on a bill will facilitate using mental math to add and subtract money in the next two chapters.

Material

- Coins and bills
- Store Cards (see materials list on p. ix)

(1) Review: Money

Discussion

Concept page 66

35 dollars **25** cents
67 dollars **25** cents
75 dollars **40** cents
32 dollars **75** cents

This should be a review of what your student learned in *Primary Mathematics* 2B. If not, point out that when we have both dollars and cents, we write a dollar sign, the total number of dollars, a dot, and then the total number of cents. We do not write a cent sign after the cents if we are using a dollar sign and a dot for money amounts. We use the cent sign only for cents.

Activity

If your student has not had much practice counting money, give her some practice before doing the tasks on page 67 of the text. You can use the following procedure. Have her write the amounts some of the times.

1. Different amounts of quarters, then dimes, then nickels up to a dollar.
2. 1 quarter and some nickels, and then 1 quarter, 1 nickel, and some dimes. Point out that it is easy to count on 25 cents to 30 cents using the the nickel before counting on with the dimes. Then let her practice with other sets of coins up to a dollar. Remind your student that we can write the amounts as cents (e.g. 67¢) or as dollars and cents, using 0 for dollars (e.g. $0.67).
3. A set of bills. Point out that it is easier to count the money if we make groups of the same denomination and start with the largest.
4. A set of coins greater than a dollar. Point out that when the amount of money is more than a dollar, we give the amount of money as dollars and cents, even when the money is all coins.
5. A set of bills and coins, including cases where the coins are more than a dollar.

Discussion

Tasks 1-5, p. 67

1. (a) **74** dollars **40** cents = **$74.40**
   (b) **7** dollars **61** cents = **$7.61**
2. (a) 125¢
   (b) $1.70
3. (a) 30¢
   (b) 195¢
   (c) 405¢
4. (a) $0.85 (b) $1.60 (c) $3.45
5. (a) $0.30 (b) $0.45

Activity

Write an amount of money, such as $401.32, in words: "four hundred one dollars and thirty-two cents." Point out that we write "dollars" after the dollar amount, the word "and" before the cent amount and then the word "cents" even though the amounts written with numerals have only a dollar sign.

Have your student write some amounts up to $10,000 in words.

Workbook

Exercise 1, pp. 67-68 (Answers p. 73)

Reinforcement

Extra Practice, Unit 9, Exercise 1, pp. 151-152

(2) Make change, practice

Discussion

Have available coins and bills. You can use pieces of paper with dollar amounts, or even the bills from a game such as Monopoly™ if you don't have play money or enough real bills.

Tell your student you want to buy something that costs 13¢ and pay for it with $1. Discuss ways to make change for a dollar, illustrating with coins. We can count up from 13¢ to 15¢ with 2 pennies, then to 25¢ with a dime, then to 50¢, 75¢, and $1 with quarters. Ask him how you would make change if you had no dimes or quarters. Repeat with a some amounts less than a dollar.

Write down the cost of an item and discuss ways to make change, using coins and bills, for various amounts paid with a $5, $10, $20, $50, or $100 bill.

Practice

Practice A, p. 68

1. (a) 20¢ (b) 65¢ (c) 700¢
   (d) 205¢ (e) 560¢ (f) 395¢
2. (a) $0.05 (b) $0.60 (c) $4.00
   (d) $2.10 (e) $8.55 (f) $3.05
3. (a) 70¢ (b) 55¢
   (c) 60¢ ($0.60) (d) 35¢ ($0.35)
4. (a) $1.50
   (b) $0.60
   (c) $2.10
5. $4.70

Reinforcement

Play store. Tag some items with a cost, or use the Store Cards. Your student can play cashier, or can select items to buy. If she is the buyer, give her a specific amount of money under $100. She will have to estimate to select items that do not exceed how much money she has.

Game

Material: Store Cards, bills and coins.

Procedure: Shuffle cards and place face-down. Give each player four $100 bills, and put the coins and smaller bills in the center of the playing area. Each player takes turns turning over a card and buying the item, putting money in the center and retrieving change. The first player to use up all his money or not have enough to buy the last item wins.

Enrichment

Ask your student for the number of

⇒ Quarters in 50 dollars.

⇒ Dimes in 50 dollars.

⇒ Nickels in 50 dollars.

⇒ Dollars in 23 quarters.

⇒ Dollars in 42 dimes.

50 dollars = ____ quarters
1 dollar: 4 quarters
50 dollars: 4 x 50 = 200 quarters

50 dollars = ____ dimes
1 dollar: 10 dimes
50 dollars: 10 x 50 = 500 dimes

50 dollars = ____ nickels
1 dollar: 20 nickels
50 dollars: 20 x 50 = 1000 nickels

23 quarters = ____ dollars
23 ÷ 4 = 5 R 3
$5 with 3 quarters left over.

42 dimes = ____ dollars
42 ÷ 10 = 4 R 2
$4 with 2 dimes left over

Test

Tests, Unit 9, 1A and 1B, pp. 105-111

Chapter 2 – Addition

Objectives

- Add money within $100 mentally by first adding dollars and then cents.
- Add money within $100 using the addition algorithm.
- Add money within $100 using mental math strategies.
- Solve word problems involving the addition of money.

Notes

Students learned various methods for adding money within $10 in *Primary Mathematics* 2A. In this chapter, your student will learn the following strategies for adding money within $100.

⇒ Add cents by first making a whole number of dollars. This method can be used when the cents add to more than a dollar, particularly when it is easy to see what needs to be added to one set of money to make a whole dollar, and what remains when this amount is subtracted from the other set of money.

$46.25 + $0.85 = $47.10
(85¢ split into 75¢ and 10¢)

⇒ Add the dollars, and then add the cents.

$46.25 + $15.85 = ?

$46.25 —(+$15)→ $61.25 —(+85¢)→ $62.10

$46.25 + $15.85 = $62.10

⇒ Use the formal algorithm for addition.
Write the problem vertically, aligning the dots (decimals) and add using the same methods as with whole numbers. This method should be used when the problem cannot be easily solved mentally.

```
   $44.85         1 1 1
+  $38.55          4485
   $83.40  ←     + 3855
                   8340
```

⇒ Add a whole number of dollars, and then subtract the difference. This method can be used when the cents in one of the amounts being added are close to 100.

$46.25 + $2.95 = ?

$46.25 —(+$3)→ $49.25 —(−5¢)→ $49.20

$46.25 + $2.95 = $49.20

The problems at this level in the textbook and workbook will generally use multiples of 5 for the cents to facilitate mental calculation.

Material

- Bills and coins
- Store Cards
- Mental Math 9 (Appendix)

(1) Add Money

Discussion

Concept page 69

We can add money by adding the dollar amounts and cent amounts separately.

$19.75
$19.75

Tasks 1-2, p. 70

1(a-c): These should be easy to do mentally.

1(d-f): Your student should recognize that the cents add up to a dollar. So the total dollar amount is increased by 1 dollar.

2(a-d): Your student should add by determining how many cents are needed to make the next dollar, and taking that from the second set of cents.

2e: Discuss two strategies for this task.

⇒ We can make $1 with the 95¢, using the 70¢, leaving 65¢.

⇒ Since 95¢ is almost $1, we can add $1 and then subtract 5¢.

1. (a) $1.70 (b) $14.85 (c) $38.75
   (d) $3.00 (e) $26.00 (f) $34.00

2. (a) $3.05
   (b) $3.30
   (c) $6.10
   (d) $17.40
   (e) $25.65

$24.70 + 95¢ = ?

$24.70 + 95¢ = $25.65
$24.65 5¢

or

$24.70 —(+$1)→ $25.70 —(−5¢)→ $25.65

$24.70 + 95¢ = $25.65

Task 3, p. 70

If your student is good at mental math, he can do these without writing down the intermediate step. Otherwise, have him write down the intermediate step.

3. (a) $29.70; $29.80
   $29.80
   (b) $36.65; $37.00
   $37.00
   (c) $35.80; $36.20
   $36.20
   (d) $34.70; $35.20
   $35.20

Top of p. 71

Remind your student that we can also add money by converting to cents, rewriting the problem with one amount above the other and adding the usual way, starting with the ones. Then we convert back to dollars and cents.

Practice

Tasks 4-5, pp. 70-71

Allow your student to use her choice of methods, the standard addition algorithm, mental math or a combination.

4. (a) $20.85 (b) $38.00
   (c) $57.20 (d) $39.30
   (e) $60.50 (f) $55.10

5. (a) $56.70
   (b) $38.00
   (c) $74.30
   (d) $69.20
   (e) $61.10
   (f) $61.75

Workbook

Exercise 2, pp. 69-71 (Answers p. 73)

Enrichment

Mental Math 9

(2) Word problems

Discussion

Tasks 6-8, pp. 71-72

You can let your student see the bar models in the text for tasks 7 and 8, and discuss how they relate to the problem. Or you can write the problem on a white board and guide your student in drawing the models and solving the problems.

7: There are two parts, the amount he paid and the amount he had left, and we need to find the total.

8: We are comparing two amounts. We are given the smaller amount, how much less it is than the larger amount, and need to find the larger amount.

6. $9.75
   $9.75
7. $36.55
   $36.55
8. $9.10
   $9.10

Workbook

Exercise 3, pp. 72-73 (Answers p. 73)

Reinforcement

Extra Practice, Unit 9, Exercise 2, pp. 153-158

Game

Material: Store cards

Procedure: Shuffle cards and place face-down. Players each turn over two cards and add the amounts. The player with the greatest sum gets all the cards. The player with the most cards at the end wins.

Enrichment

During shopping trips, help your student keep a running estimate of the total cost of items you put in the cart. Let him see how close he is with his estimate at checkout.

Test

Tests, Unit 9, 2A and 2B, pp. 113-118

Chapter 3 – Subtraction

Objectives

- Subtract money within \$100 by first subtracting dollars and then cents.
- Subtract money within \$100 using the subtraction algorithm.
- Subtract money within \$100 using mental math strategies.
- Solve word problems involving the subtraction of money.

Notes

Students learned various strategies for subtracting money within \$10 in *Primary Mathematics* 2A. In this chapter, your student will learn the following strategies for subtracting money within \$100.

⇒ If there are not enough cents to subtract from, then subtract the cents from one of the dollars. This method can be used when it is easy to "make 100" with the cents, and add this amount to the other set of cents.

$36.25 – $0.75 = ?
$35.25 $1
$36.25 – $0.75 = $35.25 + $0.25
= $35.50

⇒ Subtract the dollars, and then subtract the cents.

$61.65 – $23.50 = ?
$61.65 —(–$23)→ $38.65 —(–50¢)→ $38.15
$61.65 – $23.50 = $38.15

⇒ Use the formal algorithm for subtraction.
Write the problem vertically, aligning the dots (decimals) and subtract using the same methods as with whole numbers.

```
  $7 1. 0 5          6 10 9
–  $3 3. 6 8          7 1 0 15
  $3 7. 3 7   ←      + 3 3 6 8
                       3 7 3 7
```

⇒ Subtract a whole number of dollars, and then add the difference. This method can be used when the cents in the amount being subtracted are close to 100.

$64.25 – $2.95 = ?
$64.25 —(–$3)→ $61.25 —(+5¢)→ $61.30
$64.25 – $2.95 = $61.30

⇒ Subtract from dollars that are multiple of tens by subtracting dollars from one less and then subtracting cents from \$1. Thus \$10 is \$9 and \$1, and \$100 is \$99 and 1, for example.

$100 – $42.75 = ?
$99 $1
$99 – $42 = $57
$1 – $0.75 = $0.25
$100 – $42.75 = $57.25

The problems at this level in the textbook and workbook will generally use multiples of 5 for the cents to facilitate mental calculation.

Material

- Bills and coins
- Store Cards
- Mental Math 10-11 (Appendix)

(1) Subtract Money

Discussion

Concept page 73

We can subtract money by subtracting the dollar amounts and cent amounts separately.

$31.20
$31.20

Tasks 1-2, p. 74

Go through these problems with your student. She should be able to do these mentally.

2: Have your student use the answer to 2(a) to solve 2(b). Then, in 2(c) and 2(d), guide your student in subtracting the cents from the dollar. Write a number bond, and show the steps, if needed.

1. (a) $2.40 (b) $8.45 (c) $35.40
   (d) $1.25 (e) $6.05 (f) $46.45

2. (a) 40¢ (b) 70¢
   (c) 90¢ (d) 55¢

$1.25 – $0.35 = ?

$0.25 $1 – $0.35 = $0.65

$0.25 + $0.65 = $0.90

Task 3, p. 74

These are meant to be solved mentally by subtracting the cents from a dollar and then adding the difference.

Point out that by subtracting from a dollar, we subtract a dollar and then add in the difference. This is easy to do when the amount we are subtracting is close to a dollar even if it does not end in 5 or 0.

Have your student solve $24.78 – 97¢.

3. (a) $2.40
   (b) $13.75
   (c) $45.80
   (d) $31.20

$24.78 – 97¢ = ?

$24.78 —(–$1)→ $23.78 —(+3¢)→ $23.81

$24.78 – 97¢ = $23.81

Task 4, p. 74

If your student is good at mental math, she can do these without writing down the intermediate step. Otherwise, have her write down the intermediate step.

4. (a) $12.80; $12.20
   $12.20
   (b) $27.70; $27.50
   $27.50
   (c) $17.20; $16.70
   $16.70

Top of p. 75

Remind your student that we can also subtract money by converting to cents, rewriting the problem with one amount above the other and subtracting the usual way, starting with the ones, and then converting back to dollars and cents.

Practice

Tasks 5-7, pp. 74-75

Allow your student to use her choice of methods, the standard subtraction algorithm, mental math, or a combination.

5. (a) \$35.50 (b) \$34.85
   (c) \$45.40 (d) \$52.80
   (e) \$9.70 (f) \$27.25

6. (a) \$21.70 (b) \$25.50
   (c) \$56.80 (d) \$16.80
   (e) \$41.45 (f) \$29.90

7. (a) 5785 (b) \$57.85
   (c) 3170 (d) \$31.70

Discussion

Task 8, p. 75

Discuss 8(a) and 8(b), and then let your student practice with the rest of the problems in this task.

8(a): Point out that since it is easy to subtract 70 cents from a dollar, we can split \$10 into \$9 and 100¢, and subtract dollars from dollars and cents from cents.

8(b): We can split \$30 into \$29 and 100¢, and subtract dollars from dollars and cents from cents.

8. (a) \$5.30 (b) \$22.80 (c) \$41.75
   (d) \$26.20 (e) \$47.10 (f) \$60.55

\$10 – \$4.70
/\
\$9 100¢

\$9 – \$4 = \$5
100¢ – 70¢ = 30¢

\$10 – \$4.70 = \$5.30

\$30 – \$7.20
/\
\$29 100¢

\$29 – \$7 = \$22
100¢ – 20¢ = 80¢

\$30 – \$7.20 = \$22.80

Workbook

Exercise 4, pp. 74-76 (Answers p. 73)

Enrichment

Mental Math 10

(2) Word problems

Discussion

Tasks 9-11, pp. 75-76

9. \$4.35 \$4.35
10. \$5.40 \$5.40
11. \$9.50 \$9.50

You can let your student see the bar models in the text for tasks 10 and 11, and discuss how they relate to the problem. Or you can write the problem on a white board and guide your student in drawing the models and solving the problems.

10: There are two parts, the cost of the umbrella and the amount of money left. We are given the total and one part, the amount left. We need to find the other part.

11: The two parts are how much money Jim has and how much he still needs. The total is the cost of the watch. Note that this problem and the previous one both start in the same way, but in the previous problem the amount Mei had was the whole, where in this problem the amount Jim has is a part.

Workbook

Exercise 5, problems 1-4, pp. 77-78 (Answers p. 74)

Reinforcement

Extra Practice, Unit 9, Exercise 3, pp. 159-162

Game

Material: Store cards

Procedure: Shuffle cards and place face-down. Players each turn over two cards and subtract the amounts. The player with the smallest difference gets all the cards. The player with the most cards at the end wins.

(3) Practice

Practice

Practice B, p. 77

Reinforcement

Use the Store Cards. Set out some of them.

Determine the sum for the amounts on two of the cards and tell your student that sum. Your student must then find the two cards that give that sum.

Determine the difference between the amounts on two of the cards and tell your student that difference. Your student must then find the two cards that give that difference.

Workbook

Exercise 5, problems 5-8, pp. 79-80 (Answers p. 74)

Test

Tests, Unit 9, 3A and 3B, pp. 119-125

Enrichment

Mental Math 11

1. (a) \$39.70 (b) \$100.00
   (c) \$75.70 (d) \$40.30
   (e) \$91.65 (f) \$91.00
2. (a) \$21.35 (b) \$36.05
   (c) \$21.60 (d) \$23.75
   (e) \$45.25 (f) \$49.15
3. \$24.60 \$76.40

   ?

   \$24.60 + \$76.40 = \$101.00
   She had **\$101** at first.
4. 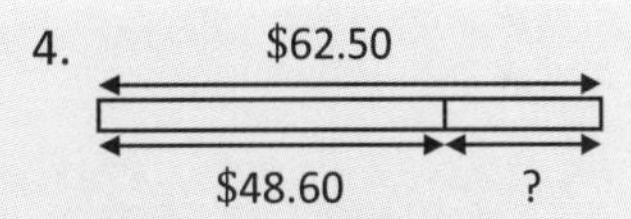

   \$62.50 – \$48.60 = \$13.90
   He needs **\$13.60**.
5. \$16.80 + \$5.60 = \$22.40
   The airplane costs **\$22.40**.
6. \$10 – \$6.95 = \$3.05
   His lunch cost **\$3.05**.
7. \$42.50 – \$16.85 = \$25.65
   The skirt costs **\$25.65**.
8. Cost of pen and book:
   \$6.80 + \$13.20 = \$20
   \$40.50 – \$20.00 = \$20.50
   She had **\$20.50** left.

Chapter 4 – Multiplication and Division

Objectives

- Multiply and divide money by a 1-digit whole number.
- Estimate to check the reasonableness of the answer.
- Apply mental math strategies to multiplication and division of money where appropriate.
- Solve word problems involving multiplication or division of money by a 1-digit whole number.

Notes

In this chapter, students will learn to multiply and divide money by a 1-digit whole number.

The easiest method to use is to convert the dollars to cents, multiply or divide, and then convert back. It is not necessary for students to write the problem as cents; they can use the dollar and cent notation if they are careful to place the dot (decimal) correctly.

We can also multiply or divide the dollars and cents separately. This can be done if it is easy to apply mental math strategies. For example, $4.50 x 6 can be done mentally because it is easy to calculate $4 x 6 ($24) and 50¢ x 6 ($3) and add them together. $35.25 ÷ 5 can be easily be done mentally, since $35 divides evenly by 5. It is not as easy to use mental math strategies if dividing the dollar amount gives a remainder.

$5.69 x 6 = ?

```
  4 5
  5 6 9
x     6
3 4 1 4
```

$5.69 x 6 = $34.14

$34.56 ÷ 6 = ?

```
     5 7 6
6 ) 3 4 5 6
    3 0
      4 5
      4 2
        3 6
        3 6
```

$34.56 ÷ 6 = $5.76

$4.50 x 6 = ?

```
      $4.50
      /    \
$4 x 6    50¢ x 6
=$24      =$3
      \    /
       $27
```

$4.50 x 6 = $27

$35.25 ÷ 5 = ?

```
       $35.25
       /     \
$35 ÷ 5     25¢ ÷ 5
=$7         =5¢
       \     /
        $7.05
```

$35.25 ÷ 5 = ?

Students will also be asked to estimate their answers. This chapter is therefore a good review of concepts learned in *Primary Mathematics* 3A for multiplication and division of multi-digit numbers by a 1-digit whole number and estimation.

Although students do not learn about decimal numbers, dollars and cents written in the decimal notation are essentially decimal numbers. These chapters are therefore also good preparation for the concept of decimals and multiplying and dividing decimal numbers in *Primary Mathematics* 4B.

(1) Multiply money

Discussion

Top of concept p. 78

One way to multiply money is to convert it to cents, multiply, and then convert the answer back to dollars and cents. Point out that is it not really necessary to rewrite the problem without the dot, as long as we remember to put the dot back into the answer.

$27.00
$27
$27

Ask your student how he could answer this problem mentally. We can separate dollars and cents and multiply them separately, then add the results. 4 dollars x 6 is 24 dollars. We could multiply 50¢ x 6 and get 300¢, which is $3. Another strategy is to remember that there are 2 half-dollars in a dollar, so 6 half-dollars is 3 dollars.

$4.50
/ \
$4 x 6 50¢ x 6
$24 $3
\ /
$27

Tasks 1-4, p. 79

1(a, c): Remind your student that any cent amount with 2 zeros at the end is the dollar amount without the zeros. We can simply multiply $4 by 7, or $2 by 3.

1(e): 50¢ x 4 can easily be done mentally; there are two half-dollars in a dollar, so 4 half-dollars is 2 dollars.

2: We can find the answer easily by thinking in terms of quarters. There are 4 quarters in a dollar, so 8 quarters is 2 dollars, and 6 is a dollar and a half.

3: Since ten dimes is a dollar, 20 dimes is 2 dollars. Ask your student to multiply other amounts, such as $0.10 x 25, which can easily be solved mentally. 20 dimes is 2 dollars and the remaining 5 dimes is 50 cents.

1. (a) 2800¢ (b) $28.00
(c) 600¢ (c) $60.00
(e) 200¢ (f) $2.00
(g) 490¢ (h) $4.90

2. (a) $2.00
(b) $1.50

3. (a) $1.00
(b) $2.00

4. $45
$43.75
$43.75

4: Point out that when we go shopping and want to buy more than one of a item, we can estimate the cost to find out about how much money we will need. Ask your student some questions involving estimation, such as the following:

⇒ Some toy cars cost $2.97. About how much money do you need to buy 10 of them?

Point out that stores often show prices close to the next dollar, hoping that you will only look at the dollar amount. If we do so, it might seem that we only need about $20 in the above example, but actually we need closer to $30.

Practice

Tasks 5-6, pp. 79-80

Workbook

Exercise 6, 81-82 (Answers p. 74)

5. (a) $15.44 (b) $44.65 (c) $11.06
(d) $56.32 (e) $61.86 (f) $86.40

6. 1 unit = **$8.95**
6 units = **$8.95** x 6
6 tickets will cost $**53.70**

(2) Divide money

Discussion

Bottom of concept p. 78

To divide money, we also convert to cents, divide, and then convert back to dollars and cents.

Get your student to estimate the answer. Remind her that in order to estimate the answer to a division problem we want to round to a whole number of dollars that is easy to divide by 9. We can round \$49.50 to either \$45 or \$54. The answer will be between \$5 or \$6.

> \$5.50
> \$5.50
> \$5.50

Tell your student that if the problem is already written in dollars, as in 9)\$49.50 , we don't have to rewrite it if we imagine the dot removed, and then put it back in the correct place, which will be right above where it is in the problem.

Enrichment: You may want to discuss an alternate solution. We could divide the dollars and the cents separately. If we divide the dollars, there is a remainder, which must be converted to cents and added to the cents before dividing the cents. So if there is a remainder when dividing the dollars, this method is not particularly easier than converting the whole amount to cents and dividing.

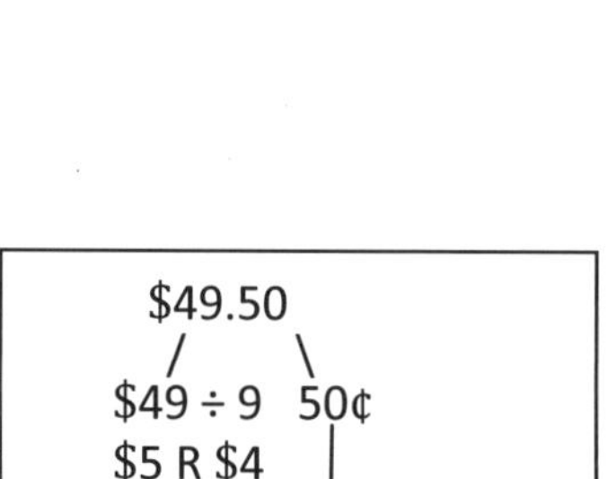

Tasks 7-9, p. 80

7: You can have your student do an estimate of the problems in the second column. Since \$36 and \$63 can be divided by 6 and 9 respectively in (b) and (f), we don't really need to convert to cents and can just divide the dollar amount. \$3.60 ÷ 6 and \$0.30 ÷ 5 will both have answers less than a dollar.

9: Note that since \$31 cannot be evenly divided by 5, we need to convert to cents and include the two zeros from the cents.

> 7. (a) 600¢ (b) \$6.00
> (c) 60¢ (c) \$0.60
> (e) 700¢ (f) \$7.00
> (g) 6¢ (h) \$0.06
>
> 8. \$6
> \$6.22
> \$6.22
>
> 9. 5)\$31.00 = \$ 6.20
> 30
> 10
> 10

Practice

Tasks 10-13, p. 81

> 10. (a) \$0.73 (b) \$8.26 (c) \$6.25
>
> 11. \$10.88 ÷ 8 = \$1.36
>
> 12. 4 units = \$**9.76**
> 1 unit = \$**9.76** ÷ 4
> Each child received \$**2.44**.
>
> 13. 3 units = \$7.23
> 1 unit = \$7.23 ÷ 3 = \$2.41
> Tom has \$**2.41**.

Workbook

Exercise 7, pp. 83-84 (Answers p. 74)

(3) Practice

Practice

Practice C, p. 82

9: Encourage your student to use estimation, rather than find the exact cost of 1 lb, or 16 oz of cherries. To find the exact cost, he would have to find the cost of 1 oz, and then multiply that by 16.

10: Point out that the better buy is where the cost of each orange is less. We can either find the cost of 8 lb of the fruit stand oranges, or the cost of 1 lb of the oranges in a bag.

Reinforcement

Extra Practice, Unit 9, Exercise 4, pp. 163-164

Test

Tests, Unit 9, 4A and 4B, pp. 127-132

Enrichment

Use opportunities when shopping to get your student to estimate the cost of items. For example, if you see the cost of apples is $1.89 a pound, ask your student what it would cost, about, to buy 5 pounds of apple. You can also use estimation to determine what product is a better buy. For example, is a shampoo packed into a 4-pack which costs $34.19 a better purchase than buying 4 individual bottles of shampoo at $8.79 a bottle? You can point out that stores usually give the price per ounce or per unit on their cost labels, which makes it easy to see which is a better buy.

1. (a) 999¢ (b) 8345¢ (c) 67¢
2. (a) $6.52 (b) $0.19 (c) $11.09
3. (a) $10.98 (b) $28.25 (c) $89.91
   (e) $56.00 (f) $3.36 (g) $0.18
   (h) $1.20 (i) $39.08
4. (a) $2.37 (b) $0.71 (c) $1.21
   (d) $1.25 (f) $0.50 (g) $5.25
   (h) $1.99 (i) $0.64
5. $5.21 x 2 = $10.42
   Amy saved **$10.42**.
6. $9.38 ÷ 2 = $4.69
   He saves **$4.69**.
7. 5 lb: $2.95
   1 lb: $2.95 ÷ 5 = $0.59
   1 lb of flour costs **$0.59**.
8. Cost for pencils:
   $5.45 x 3 = $16.35
   Total cost:
   $16.35 + $2.95 = $19.30
   He has to spend **$19.30**.
9. There are 16 oz in a pound.
   15 oz are 3 sets of 5 oz.
   5 oz: $1
   15 oz = $1 x 3 = $3
   16 oz would cost more than that, so the cherries are not less than $3 a pound and she should **not buy** them.
   Or
   1 oz. costs 100¢ ÷ 5 = 20¢
   16 oz costs 16 x 20¢ = $3.20
   The cherries cost too much.
10. Cost of 8 lb of fruit stand oranges:
    $1.23 x 8 = $9.84
    Or
    Cost of 1 lb of the oranges in the bag:
    $10.80 ÷ 8 = $1.35
    Bagging your own oranges is a better buy.

Review 9

Review

Review 9, pp. 83-84

Workbook

Review 9 pp. 85-89 (Answers p. 75)

Test

Tests, Units 1-9, Cumulative A and B, pp. 133 -144

1. (a) 5932 (b) 6808 (c) 3600
2. (a) 999 (b) 2924 (c) 5336
3. (a) 308 (b) 657 (c) 615
4. (a) 450 (b) 136 (c) 64 r6
5. $504 ÷ $8 = 63
   He sold **63** pens.
6. $628 + $1485 + $515 = $2628
   He had **$2628** at first
7. Apples in each box: 12 + 8 = 20
   Total apples: 20 x 4 = 80
   There were **80** apples.
8. Number of boxes sold this month:
   337 + 299 = 636
   Total boxes sold: 636 + 337 = 973
   He sold **973** boxes in the two months.
9. (a) 200 ÷ 8 = 25
   She made **25** cakes.
   (b) $10 x 25 = $250
   She would receive **$250**.
10. (a) $41.00 (b) $3.25
11. (a) $30.60 (b) $25.15
12. (a) $100.00 (b) $18.95
13. (a) $63.40 (b) $9.10
14. (a) $99.90 (b) $28.30
15. (a) $14.28 (b) $71.04
16. (a) $2.31 (b) $3.89
17. $42.50 – $15.90 = $26.60
    The badminton racket is **$26.60** cheaper.
18. $43.00 – $29.95 = $13.05
    The sale price is **$13.05** cheaper.
19. (a) $64.80 ÷ 6 = $10.80
    She saved **$10.80** each week.
    (b) $82.30 – $64.80 = $17.50
    Her mother gave her **$17.50**.
20. $2.40 + $3.70 + $21.30 = $27.40
    He had **$27.40** at first.
21. (a) $5.70 x 3 = $17.10
    She spent **$17.10** on the chickens.
    (b) Cost of duck: $5.70 + $1.95 = $7.65
    Total spent: $7.65 + $17.10 = $24.75
    She spent **$24.75** altogether.

Workbook

Exercise 1, pp, 67-68

1. (a) $80.27
   (b) $4.30
   (c) $30.85

2. (a) $5.65
   (b) $10.08
   (c) $17.70
   (d) $90.12
   (e) $320.04
   (f) $1030.00

3. (a) eighty cents
   (b) one dollar and thirty-six cents
   (c) six dollars and forty-four cents
   (d) seven dollars and ninety-eight cents
   (e) twenty-three dollars and twenty cents
   (f) ten dollars and five cents
   (g) forty-four dollars and fifty-five cents
   (h) four hundred twelve dollars
   (i) three thousand, seven hundred nine dollars

Exercise 2, pp. 69-71

1. (a) $11.90 $11.95   (b) $9.85 $10.25
   (c) $34.35 $35.10   (d) $66.35 $67.20

2. (a) $31.30   (b) $35.10
   (c) $69.15   (d) $53.40

3. (a) 1000 $10.00
   (b) 3115 $31.15
   (c) 9485 $94.85
   (d) 10,000 $100.00

4. A. $4.95   E. $16.35   G. $17.60
   I. $80.29   L. $93.73   N. $85.20
   R. $43.85   S. $53.14   T. $73.90
   TRIANGLES

Exercise 3, pp. 72-73

1. (a) $10.70 + $ 1.20 = $11.90
   (b) $ 2.75 + $ 3.15 = $ 5.90
   (c) $17.80 + $ 2.75 = $20.55
   (d) $ 0.95 + $12.90 = $13.85
   (e) $ 2.75 + $ 1.20 + $ 0.95 = $ 4.90
   (f) $17.80 + $12.90 + $10.70 = $41.40

2. $92.65 + $79.80 = $172.45
   He now has **$172.45** in the bank.

3. $62.05 + $98.83 = $160.88
   The cooker costs **$160.88**.

4.
$80.80 $98.70
radio
television
?

$80.80 + $98.70= $179.50
The television set costs **$179.50**.

Exercise 4, pp. 74-76

1. (a) $2.50 $2.45   (b) $2.35 $1.55
   (c) $12.00 $11.40   (d) $17.05 $16.60

2. (a) $5.15   (b) $4.05
   (c) $53.60   (d) $53.55

3. (a) 5250 $52.50
   (b) 2525 $25.25
   (c) 4520 $45.20
   (d) 3515 $35.15

4. $37.20 $25.50 $16.15
   $63.45 $16.55 $7.01
   $35.95 $15.85 $9.90
   BADMINTON

Workbook

Exercise 5, pp. 77-80

1. (a) \$5 – \$2.75 = **\$2.25**
   (b) \$3 – \$1.65 = **\$1.35**
   (c) \$20.50 – \$19.50 = **\$1.00**
   (d) \$50.25 – \$24.80 = **\$25.45**
   (e) \$72 – \$69.90 = **\$2.10**
2. \$60.25 – \$16.90 = \$43.35
   The watch cost **\$43.35**.
3. 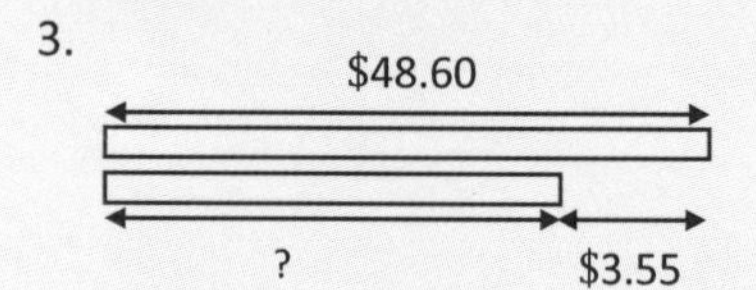

   \$48.60 – \$3.55 = \$45.05
   She spent **\$45.05**.
4. \$70 – \$47.95 = \$22.05
   The watch is **\$22.05** cheaper at the sale.
5. Total spent:
   \$24.95 + \$9.50 = \$34.45
   Change:
   \$50 – \$34.45 = \$15.55
   She received **\$15.55** change.
6. 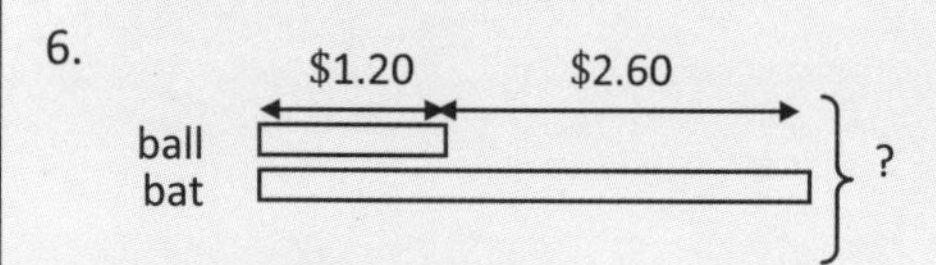

   Cost of bat:
   \$1.20 + \$2.60 = \$3.80
   Total cost:
   \$3.80 + \$1.20 = \$5.00
   She spent **\$5.00**.
7. 

   Amount spent:
   \$10.00 – \$2.05 = \$7.95
   Cost of pen:
   \$7.95 – \$7.35 = \$0.60
   The pen costs **\$0.60**.
8. 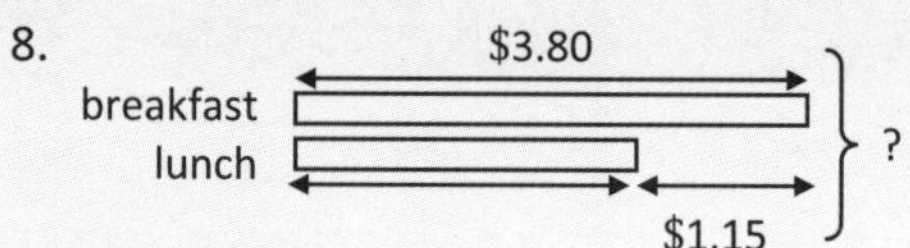

   Cost of lunch:
   \$3.80 – \$1.15 = \$2.65
   Total money:
   \$3.80 + \$2.65 = \$6.45
   He had **\$6.45** at first.

Exercise 6, pp. 81-82

1. (a) \$15.12
   (b) \$43.20
   (c) \$1.56
   (d) \$32.84
2. \$1.58 x 8 = \$12.64
   She paid **\$12.64**.
3. \$3.00 x 5 = \$15.00
   The total price of 5 tickets is **\$15.00**.
4. \$1.58 x 7 = \$11.06
   The total cost is **\$11.06**.
5. \$9.80 x 5 = \$49.00
   He earned **\$49**.

Exercise 7, pp. 83-84

1. (a) \$1.78
   (b) \$0.68
   (c) \$1.13
   (d) \$4.15
2. \$9.21 ÷ 3 = \$3.07
   The price for 1 box of bread sticks is **\$3.07**.
3. \$7.55 ÷ 5 = \$1.51
   Each sticker costs **\$1.51**.
4. \$1.44 ÷ 9 = \$0.16
   Each pencil costs **\$0.16**.

Workbook

Review 9, pp. 85-89

1. 3263
   2030
2. (a) 6553
   (b) 3944
   (c) 9107
   (d) 4590
3. 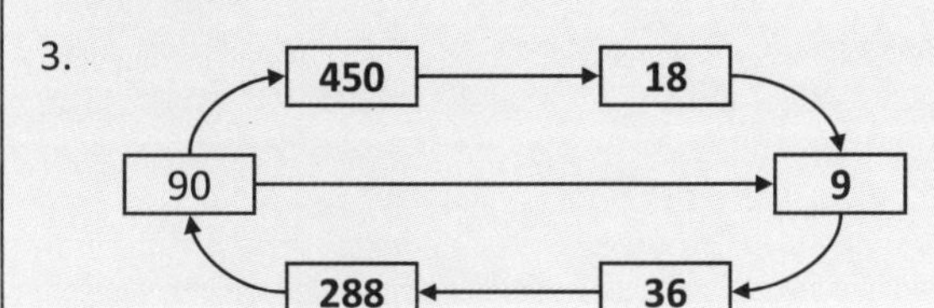

4. (a) 1 kg 405 g
   (b) 5 m 85 cm
   (c) 3 km 100 m
   (d) 8 kg 540 g
5. (a) 1 mi < 5820 ft (b) 1 in. > 1 cm
   (c) 1 yd < 1 mi (d) 1 yd = 36 in.
   (e) 1 kg > 1 lb (f) 20 oz < 21 oz
   (g) 6 c < 7 c (h) 17 c < 18 c
   (i) 33 c > 32 c (j) 30 qt = 30 qt
6. 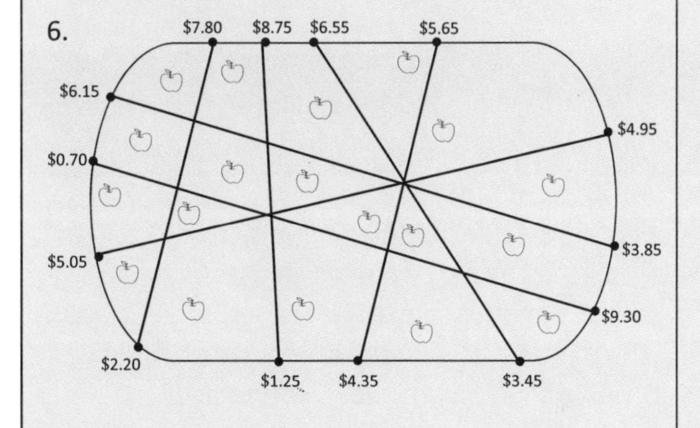

7. 2 lb 8 oz + 2 lb 9 oz = 5 lb 1 oz
   The total weight is **5 lb 1 oz.**
8. 1 can: 256 g
   9 cans: 256 g x 9 = 2304 g.
   9 cans weigh 2304 g or **2 kg 304 g**.
9. (a) $56 – $44 = **$12**
   (b) **Lance**
10. 4 kg: $8.48
    1 kg: $8.48 ÷ 4 = $2.12
    1 kg of fish costs **$2.12.**
11. Cost of 7 books: $12.45 x 7 = $87.15
    Total spent: $87.15 + $25.95 = $113.10
    She spent **$113.10** altogether.
12. Cost of lunches: $12.45 x 4 = $49.80
    Cost for each: $49.80 ÷ 6 = $8.30
    Each person paid **$8.30.**
13. 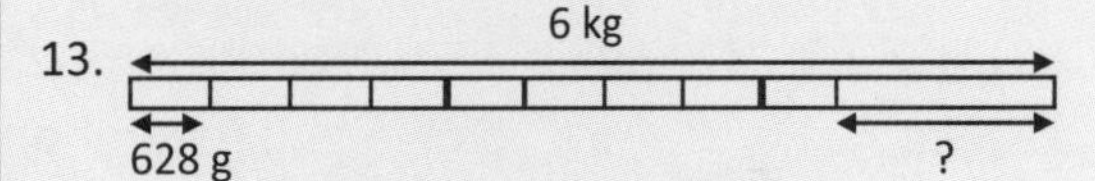

    Weight of fish: 628 g x 9 = 5652 g = 5 kg 652 g
    Weight of prawns: 6 kg – 5 kg 652 g = 348 g
    The prawns weigh 348 g.
14. Total length: 125 m + 215 m = 340 m
    Total cost: 340 m x $5 = $1700
    He paid **$1700.**

Unit 10 – Fractions

Chapter 1 – Fractions of a Whole

Objectives

- Recognize and name fractions of a whole.
- Add two fractions with the same denominator to make a whole.
- Compare and order fractions with a common numerator.
- Compare and order fractions with a common denominator.

Vocabulary

- Fraction
- Numerator
- Denominator

Notes

In *Primary Mathematics* 2B students learned to understand and write fractional notation, to find sums of fractions with the same denominators that make a whole, and to order unit fractions (fractions with 1 in the numerator). These concepts are reviewed in this chapter. The terms **numerator** and **denominator** are formally introduced here, and students will learn how to compare and order fractions with a common numerator or denominator.

Although your student should eventually learn the terms numerator and denominator, do not let math vocabulary interfere with understanding. Your student may be better able to focus on the mathematics when the informal terms **top** and **bottom** are used instead or in addition to the formal terms

$$\frac{\text{numerator}}{\text{denominator}} \quad \frac{3}{5} \quad \frac{\text{top}}{\text{bottom}}$$

Natural numbers count objects. The object being counted, such as apples or centimeters, is the denomination of the number. Fractions count *parts* of objects. The part from which the fraction is taken is called the *whole*. The whole for a fraction is like the denomination of a number. When we take three fourths of an apple, the whole is the apple, and the parts being counted are fourths. When we take three fourths of 12, the whole is 12, and the parts being counted are each a fourth of 12.

$\frac{1}{4}$ represents 1 out of 4 equal parts of the whole. $\frac{3}{4}$ represents 3 out of 4 equal parts of the whole. The top number (numerator) counts the number of fractional parts or units. The bottom number (denominator) tells us how many fractional parts or units make up the whole. The denominator also indicates the size of the part; the larger the denominator the smaller the size since the whole is divided up into more parts.

When comparing or doing mathematical operations on fractions, each fraction must have the same whole. $\frac{1}{2}$ of an inch is not greater than $\frac{1}{4}$ of a foot. We cannot add $\frac{1}{4}$ and $\frac{3}{4}$ to make a whole if they represent $\frac{1}{4}$ of a centimeter and $\frac{3}{4}$ of a meter. $\frac{1}{2}$ of 12 is not the same as $\frac{1}{2}$ of 6.

Make sure your student understands that a fraction is a *single* number, even though it is written as one number on top of the other. The top is the number of parts, the bottom the size of the parts. If your student has trouble with this, use measurement or a number line. When we draw a number line, we go from 0 to 1. On a meter stick, we start at 0 meters, and then show 1 meter. A point half way along the meter stick is $\frac{1}{2}$ of a meter. A point half way between 1 and 2 on a number line is a single point.

0 $\frac{1}{2}$ 1

Since the numerator counts the number of parts, it is easy to compare fractions with the same denominator. $\frac{2}{5}$ is smaller than $\frac{4}{5}$. The denominators are the same, so the size of the parts is the same. The fraction with the smaller numerator is therefore the smaller number of parts.

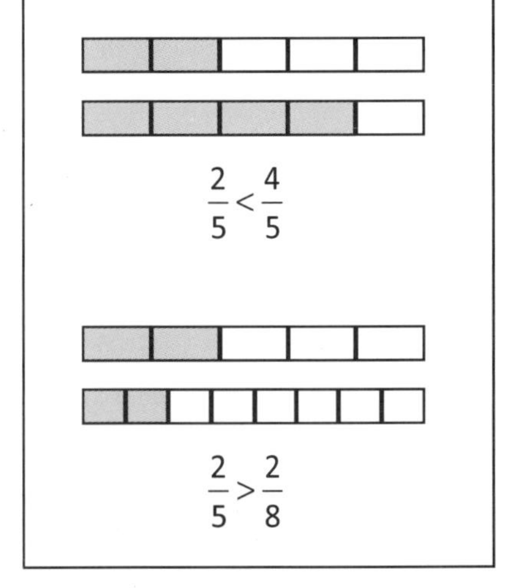

We can also easily compare fractions with the same numerator. $\frac{2}{5}$ is greater than $\frac{2}{8}$ of the same whole because the size of the parts in $\frac{2}{5}$ is greater than the size of the parts for $\frac{2}{8}$.

If there are the same number of parts, then the fraction with the smaller denominator is greater than the fraction with the larger denominator. Comparing fractions with a common numerator and different denominators help students better understand the role of the denominator in fractions.

Material

- Fraction bars, labeled (appendix p. a9)
- Fraction bars, unlabeled (appendix p. a10, or draw them)

(1) Review: Fractions

Activity

Write a fraction, such as $\frac{3}{4}$, and give your student an object that can be cut up or divided into parts, such as an apple, a lump of clay, or a strip of paper. Ask him to use the object to explain what three fourths means. Tell your student that with fractions, sometimes a fourth is called a quarter, so three fourths is sometimes called three quarters.

Discussion

Concept page 85

Task 1, p. 86

1(a): Point out that the top number in the fraction $\frac{2}{5}$ counts the number of parts, and the bottom number is how many parts the bar has been divided into.

1(c): Write the equation $\frac{2}{5}+\frac{3}{5}=1$ and have your student read it. Tell your student that when we add two fifths and three fifths, we have a whole. Write the equation $\frac{1}{5}+\frac{1}{5}+\frac{1}{5}+\frac{1}{5}+\frac{1}{5}=1$, one fifth at a time, getting your student to tell you when you have enough fifths to make a whole. There are five $\frac{1}{5}$'s in one whole.

1. (a) **2** out of the **5** equal parts.
   2 fifths
   (b) **3** out of the **5** equal parts.
   3 fifths
   (c) **5** fifths
   $1 = \frac{\mathbf{5}}{5}$

$\frac{2}{5}+\frac{3}{5}=1$ $\quad$ $\frac{1}{5}+\frac{1}{5}+\frac{1}{5}+\frac{1}{5}+\frac{1}{5}=1$

Task 2, p. 86

2(c): Ask your student to write an addition equation for this statement. Get her to write other addition equations for making a whole with eighths.

2. (a) $\frac{\mathbf{5}}{\mathbf{8}}$ of the bar is not shaded.
   (b) **8** eighths; $1 = \frac{\mathbf{8}}{8}$
   (c) $\frac{3}{8}$ and $\frac{\mathbf{5}}{\mathbf{8}}$ make 1 whole.

$\frac{3}{8}+\frac{5}{8}=1$ $\quad$ $\frac{2}{8}+\frac{2}{8}+\frac{4}{8}=1$

Activity

Write the two fractions five fifths and eight eighths and ask your student if they are equal. They are, since they are both 1 whole. Ask your student to write some other fractions equal to these two.

$\frac{5}{5}=\frac{8}{8}$ $\quad$ $\frac{5}{5}=\frac{8}{8}=\frac{7}{7}=\frac{12}{12}...$

Discussion

Task 3, p. 87

Have your student also read the fractions out loud. You can ask your student for the fraction that is unshaded as well.

3. (a) $\frac{1}{5}$ (b) $\frac{1}{6}$
   (c) $\frac{1}{12}$ (d) $\frac{2}{3}$
   (e) $\frac{2}{5}$ (f) $\frac{5}{6}$
   (g) $\frac{7}{8}$ (h) $\frac{7}{10}$

Workbook

Exercise 1, pp. 90-95 (Answers p. 104)

Reinforcement

Extra Practice, Unit 10, Exercise 1A, pp. 175-176

(2) Compare fractions

Discussion

Task 4, p. 88

4. (a) 2 is the numerator, 5 is the denominator
   (b) 4 is the numerator, 10 is the denominator
   (c) 6 is the numerator, 7 is the denominator
   (d) 6 is the numerator, 9 is the denominator.

Tell your student that the top number on a fraction is called the numerator and the bottom part is called the denominator. Ask him what each stands for. The numerator is the number of parts being counted. The denominator is the number of parts that make up the entire whole. It can help to tell the two names, numerator and denominator, apart by thinking of the word "number" which is close to the word "numerator" which is the "number" of parts we are counting.

Activity

$\frac{1}{12}, \frac{1}{11}, \frac{1}{10}, \frac{1}{9}, \frac{1}{8}, \frac{1}{7}, \frac{1}{6}, \frac{1}{5}, \frac{1}{4}, \frac{1}{3}, \frac{1}{2}, \frac{1}{1}$

Use the fraction bars from the appendix that have the fractions labeled. Have your student color one unit of each and cut it apart from the rest. Mix them up and have your student put them in order, starting with the smallest. Then have your student look at the denominator (bottom) of each. Ask why the smallest fraction has the largest denominator. The denominator tells us how many parts make the whole, and the more parts there are the smaller each part must be. You can include one over one, which is a whole, and is the greatest.

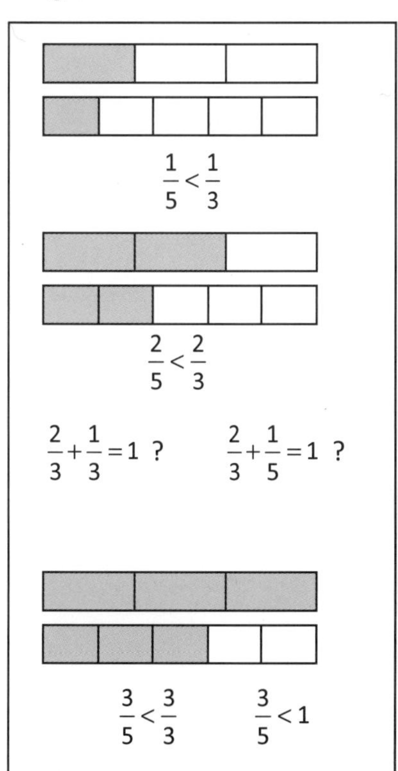

Use the unlabeled fraction bars and cut out the strips showing thirds and fifths, or simply draw the bars. Have your student color the first part of each and ask which one is greater. Write the inequality. One third is greater than one fifth. We can look at the denominator to compare the fractions; the larger denominator means the part is smaller.

Now have your student color the next part of each bar. Write the two fractions, and ask which is greater. Two thirds are greater than two fifths. Two larger parts are, of course, greater than two smaller parts.

Write $\frac{2}{3} + \frac{1}{3} = 1$ and ask if it is true. It is. Then write $\frac{2}{3} + \frac{1}{5} = 1$ and ask if it is true. It is not. Ask why not? The fifth is smaller than the third, so adding it to two thirds will not fill in the rest of the whole bar. $\frac{2}{3} + \frac{1}{5} < 1$

Have your student color in another part, write the fractions, and tell you which is greater. Again, the fraction with the smaller denominator is greater, if the number of parts (numerators) for both is the same. Three thirds is the same a one whole.

Now write the two fractions two fifths and three fifths and ask your student which is greater. See if your student notices that we now have the same denominator, so the size of the parts are the same. Three fifths is greater than two fifths because there are more parts. You can draw a bar under the one that now shows three fifths and have your student color in two fifths to see that it is smaller.

$\frac{2}{5} < \frac{3}{5}$

Discussion

Tasks 5-9, pp. 88-89

Have your student explain her answers in words. Make sure she notices which parts of each pair of fractions are the same and can explain how that led to the correct answers. In tasks 5 and 6 the numerators are the same, and the larger fraction has the smaller denominator (fewer parts means each part is larger). In task 7, the denominators are the same, so for both fractions the size of the parts is the same, and the larger fraction has the larger numerator (number of parts).

5. $\frac{1}{3}$
6. $\frac{3}{4}$
7. $\frac{5}{8}$
8. $\frac{3}{10}$ is the smallest fraction.
   $\frac{3}{5}$ is the greatest fraction.
9. $\frac{3}{9}$ is the smallest fraction.
   $\frac{7}{9}$ is the greatest fraction.

Practice

Task 10, p. 89

10. (a) $\frac{1}{7}, \frac{1}{5}, \frac{1}{3}$ (b) $\frac{2}{9}, \frac{2}{7}, \frac{2}{3}$
    (c) $\frac{4}{8}, \frac{5}{8}, \frac{7}{8}$ (d) $\frac{4}{12}, \frac{5}{12}, \frac{9}{12}$

Workbook

Exercises 2-3, pp. 96-99 (Answers p. 105)

Reinforcement

Extra Practice, Unit 10, Exercise 1B, pp. 177-178

(3) Practice

Practice

Practice A, p. 90

Activity

Ask your student to put the fractions shown at the right in order, and then explain his answer.

He has not yet learned how to compare fractions with different numerators as well as denominators without drawings yet. However, he can first compare the two fractions with the same numerator, and then the two with the same denominator.

Similarly, have your student put the fractions shown here at the right in order. He should realize that six sixths must be the greatest.

$\frac{3}{5}, \frac{3}{7}, \frac{4}{5}$?

$\frac{3}{7} < \frac{3}{5} < \frac{4}{5}$

$\frac{5}{8}, \frac{6}{6}, \frac{3}{9}, \frac{3}{8}$?

$\frac{3}{9} < \frac{3}{8} < \frac{5}{8} < \frac{6}{6}$

1. (a) $\frac{3}{4}$
   (b) $\frac{7}{10}$
   (c) $\frac{5}{12}$
2. (a) 2 (b) 6 (c) 9
3. (a) 8 (b) 9 (c) 10
4. (a) $\frac{4}{5}$ (b) $\frac{1}{4}$ (c) $\frac{3}{5}$
5. (a) $\frac{3}{10}$ (b) $\frac{1}{10}$ (c) $\frac{2}{9}$
6. (a) $\frac{5}{7}$ (b) $\frac{1}{2}$
7. (a) $\frac{1}{6}$ (b) $\frac{3}{10}$

Test

Tests, Unit 10, 1A and 1B, pp. 145-149

Enrichment

Get your student to estimate with fractional lengths. Draw some bars the same size as the whole in the fraction bars in the appendix, or the whole for commercial fraction bars you might have. Color a portion of a bar and ask him about how much is shaded. Then have him check his estimation with the fraction bars. If this is too hard, you can give him three choices and have him choose the best estimate. For example, for the drawing at the right, you can give him the choices of $\frac{4}{5}$, $\frac{1}{2}$ and $\frac{9}{12}$.

About $\frac{4}{5}$ of the bar is shaded.

Chapter 2 – Equivalent Fractions

Objectives

- Recognize and name equivalent fractions for fractions with denominators up to 12.
- Find equivalent fractions.
- Find the simplest form of a fraction.
- Compare and order fractions.

Vocabulary

- Equivalent fractions
- Simplify a fraction
- Simplest form

Notes

In this chapter, students will be introduced to the concept of equivalent fractions.

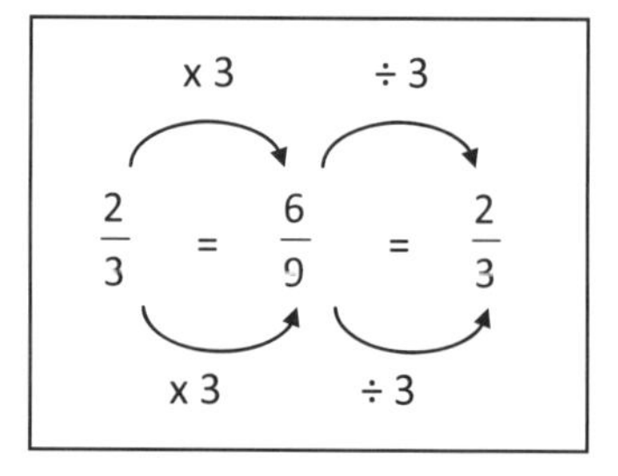

$\frac{1}{2}$, $\frac{2}{4}$, $\frac{3}{6}$, and $\frac{4}{8}$ are different names for the same fraction; they all name the same part of the whole. We can change a fraction into an equivalent fraction by multiplying or dividing both the numerator and denominator by the same number.

A fraction is *simplified* when its numerator and denominator can be divided by the same number. $\frac{6}{9}$ can be simplified to the equivalent fraction $\frac{2}{3}$ by dividing both the numerator and denominator by 3. If it is not possible to divide both the numerator and denominator by any same number (except 1), the fraction is said to be in its simplest form. Of the equivalent fractions $\frac{1}{2}$, $\frac{2}{4}$, $\frac{3}{6}$ and $\frac{4}{8}$. $\frac{1}{2}$ is the simplest form. The numerator and denominator do not have a common factor.

In this chapter, students will use equivalent fractions to compare and order fractions where neither the numerator nor the denominator are the same. To compare fractions, we need to have fractions where the denominators or numerators are the same. For example, to compare $\frac{4}{5}$ and $\frac{7}{10}$ we can rename $\frac{4}{5}$ as the equivalent fraction $\frac{8}{10}$. Now both fractions have the same denominator, and so we know $\frac{7}{10}$ is smaller than $\frac{4}{5}$ $\left(\frac{8}{10}\right)$. At this level, students will only be comparing fractions where the denominator of one fraction is a simple multiple of the denominator of the other fraction. They will compare unrelated fractions in *Primary Mathematics* 4A.

At this level, students will only look at proper fractions. Improper fractions and mixed numbers will be introduced in *Primary Mathematics* 4A.

Material

- Strips of paper
- Labeled fraction bars (appendix p. a10), 3 copies
- Appendix p. a13

(1) Understand equivalent fractions

Discussion

Concept page 91-92

Have your student actually do this task, rather than just look at the pages. To better see the equivalence, you can use three separate strips. Fold each one in half (or have her do the folding), open them up, draw a line at the fold, and color a half on each. Then fold one of them in half twice and the other in half three times. Open each up and draw lines at the folds. Ask your student to write what fraction of each is shaded. Ask her whether the shaded part is equal on each of the three strips. They are.

$\frac{1}{2}$

$\frac{2}{4}$

$\frac{4}{8}$

$\frac{1}{2} = \frac{2}{4} = \frac{4}{8}$

Tell your student that $\frac{1}{2}$, $\frac{2}{4}$, and $\frac{4}{8}$ are different ways of naming the same fraction. They are equal, even though they have different numerators and denominators. They are called equivalent fractions.

Ask your student to compare the numerator to the denominator of each of these fractions. The numerator is half of the denominator. This makes sense; one half means we are counting half of the fractional parts. Ask your student if $\frac{3}{6}$ is another name for $\frac{1}{2}$. It is. Ask him to name some other fractions that are equivalent to one half.

Activity

Draw a rectangle and draw two vertical lines to divide it into thirds. Ask your student to shade two parts and write the fraction of the rectangle that is shaded. Then, draw a horizontal line to divide the rectangle into half. Ask your student how many total parts there are, how many parts are shaded, and to write another fraction for the shaded part. Ask if the two fractions are equivalent. They are. The shaded amount does not change. Repeat with other examples as needed. The example shown at the right has a fraction with 15 in the denominator; your student should know that fractions don't stop at twelfths.

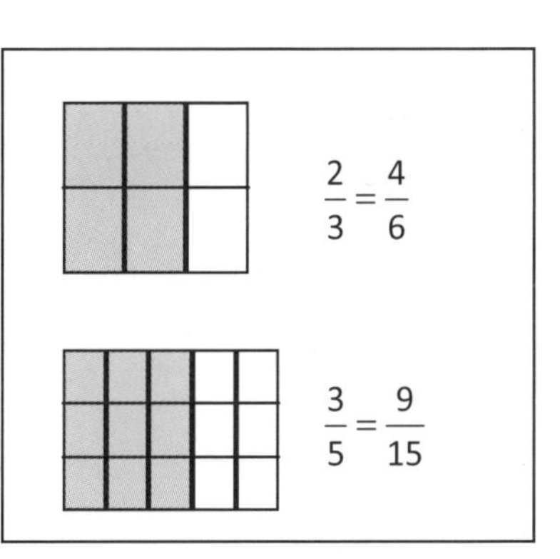

Practice

Task 1, p. 92

1(d): You can let your student use rectangles as in the activity above.

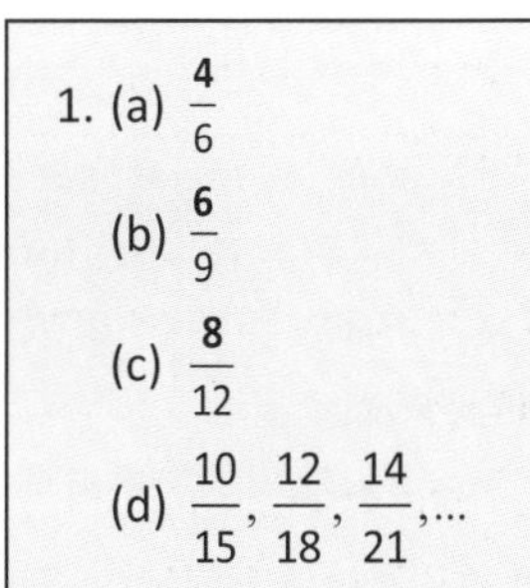

Workbook

Exercise 4, pp. 100-101 (Answers p. 106)

Reinforcement

Give your student a copy of the labeled fraction bars. You may want to cut apart the bars so they are easier to set one on top of the other to compare. Or use commercial fraction bars. Ask your student to write down as many equivalent fractions she can find.

(2) Use multiplication to find equivalent fractions

Activity

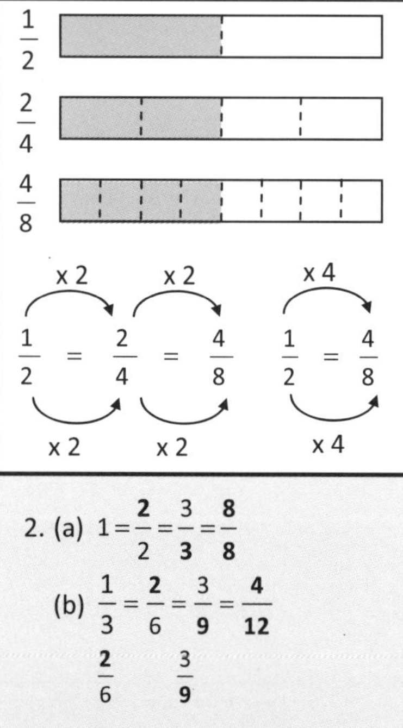

Use the strips of paper from the previous lesson or new ones. Ask your student what happens each time the strip is folded and the parts marked. Each time, the parts get half as large, and both the total number of parts and the shaded number of parts double. So each time both the numerator (top) and the denominator are doubled. You can show this with equations and arrows, as shown here at the right.

Ask your student how we can go from a half to four eighths. We can multiply both the top and bottom by 4.

Discussion

Task 2, p. 93

You can use the top half of appendix p. a13 so that your student can write the answer and then see the answer in relation to the fraction circles.

2(a): Point out that the whole has been divided so that it now has 2, 3, or 6 equal parts.

2(b): Point out that in the second fraction circle, each part is doubled. Both the top and the bottom of the fraction must be doubled to get the equivalent fraction. In splitting the parts into two, the size of each part is halved, and the shaded fraction remains the same. Ask your student how many times each part is in the third and fourth fraction circle compared to the first fraction circle. Point out that there is not a whole number we can use to go from two sixths to three ninths, or from three ninths to four twelfths, but they are still equivalent fractions and are equal to one third.

In summary, we can rename a fraction into an equivalent fraction by multiplying and dividing the top and bottom by the same number.

Practice

Task 3, p. 93

3. (a) $\frac{\mathbf{3}}{12}$ (b) $\frac{\mathbf{6}}{9}$ (c) $\frac{\mathbf{2}}{10}$

(b) $\frac{3}{\mathbf{18}}$ (b) $\frac{6}{\mathbf{10}}$ (c) $\frac{6}{\mathbf{8}}$

Your student must first find the number the numerator or denominator in the first fraction is multiplied by to get the given numerator or denominator in the second fraction. If your student has trouble with this, he may need more concrete examples. Allow him to use a copy of the labeled fraction bars in the appendix, or illustrate the problems with fraction rectangles, and then follow this practice up with additional problems until he does not need to use the fraction bars .

Workbook

Exercise 5, pp. 102-103 (Answers p. 106)

In problem 3 of this exercise, the pair $\frac{2}{6}$ and $\frac{3}{9}$ cannot be found with multiplication by a whole number, but is illustrated on the previous page by comparing both to one third.

(3) Use division to find equivalent fractions

Activity

You can use the strips of paper from the previous lessons or make new ones. Use the strips showing four eighths and one half. Write the two equivalent fractions. Ask your student how we can go from four eighths to one half. We can divide both the numerator and the denominator by 4. Point out that we are combining 4 parts into one. So 8 parts becomes 2. The size of each part is now four times larger, so the amount shaded stays the same.

$\frac{4}{8}$

$\frac{1}{2}$

$$\frac{4}{8} = \frac{1}{2} \quad (\div 4)$$

Discussion

Task 4, p. 94

You can use the bottom half of appendix p. a13 so that your student can write the answer and then see the answer in relation to the fraction circles. There is an additional problem on the page from the appendix.

Point out that two parts are combined into one part each time. There are half as many shaded parts, and half as many total parts each time, so we divide both the numerator and denominator by 2. We can skip the middle fraction circle by combining 4 parts into one.

You can use the last row of fraction circles on the appendix page to discuss the process of going from four tenths to two fifths.

4. (a) $\frac{8}{12} = \frac{\mathbf{4}}{6} = \frac{2}{\mathbf{3}}$

$\frac{\mathbf{4}}{6}$ $\frac{2}{\mathbf{3}}$

$$\frac{4}{10} = \frac{2}{5} \quad (\div 2)$$

Practice

Task 5, p. 94

If your student has difficulty determining what number she needs to divide by, allow her to use the fraction bars and then do the first problem in the workbook exercise, and then follow up with more examples until she can do the problems without the bars before doing the second problem in the exercise.

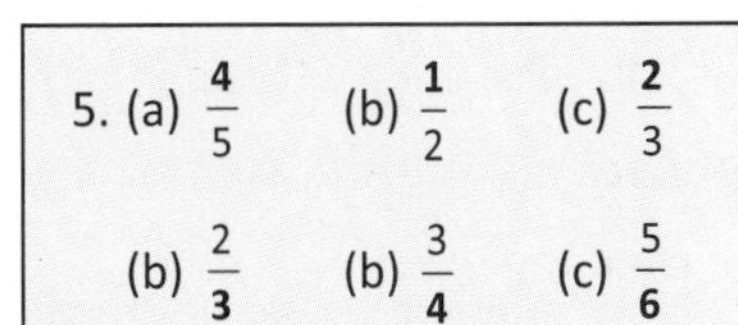

Workbook

Exercise 6, pp. 104-105 (Answers p. 106)

Reinforcement

Give your student 4 numbers and have him make up two equivalent fractions using those numbers (two of the numbers have to be the same multiples of the other two). Then give your student 6 numbers to form into equivalent fractions.

4, 10, 8, 5 $\quad \frac{8}{10} = \frac{4}{5}$

12, 4, 6, 3, 8, 9 $\quad \frac{3}{4} = \frac{6}{8} = \frac{9}{12}$

(4) Find the simplest form of a fraction

Activity

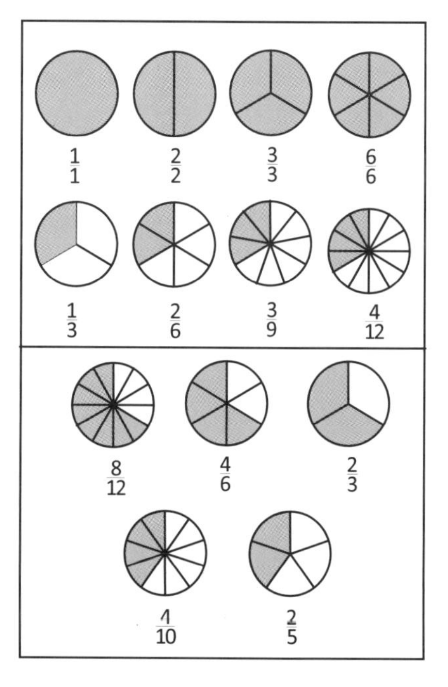

Use appendix p. a13 which your student completed in the last two lessons. Have your student look at each row and discuss the following ideas.

In the top row, we can combine parts to go from the second through fourth fraction circle to the first one, which represents 1 whole.

In the second row, we can combine 2, 3, or 4 parts to go from each of the others to the first fraction circle, which shows one third shaded.

In the third row, we combined four and then two parts into one to get the equivalent fraction two thirds, where there are two shaded parts. Can we combine parts any further? No, since if we combine the two shaded parts to make one part, there are not enough unshaded parts left to have equal parts.

In the last row, with two fifths we cannot combine shaded parts any more since we will then be left with an unequal part.

Tell your student that one, one third, two thirds, and two fifths are the **simplest form** of each of the sets of equivalent fractions. It is the form with the smallest possible number in the numerator, or top. Each time we go from one fraction to an equivalent fraction with a smaller numerator, we **simplify** the fraction.

Have your student look at the numerical fractions as ask how we could determine if a fraction is in the simplest form. Lead her to see that any time both the numerator and denominator have even numbers the fraction cannot be in its simplest form; we can still combine parts in pairs. Both the numerator and denominator can be divided by 2. For fractions where both are not even, any time the denominator can be evenly divided by the numerator we do not have the simplest form of the fraction. Also, any time the numerator and denominator can both be divided by the same number we do not have the simplest form of the fraction.

To find the simplest form, we can keep dividing the numerator and denominator by the same number until there is no whole number that both can be divided by (other than 1 which does not change the fraction).

Practice

Tasks 6-7, pp. 94-95

Workbook

Exercise 7, pp. 106-107 (Answers p. 107)

Reinforcement

Extra Practice, Unit 10, Exercise 2, pp. 181-184

6. (a) $\frac{3}{4}$ (b) $\frac{2}{4}$ (c) $\frac{1}{2}$
   $\frac{1}{2}$

7. (a) $\frac{1}{2}$ (b) $\frac{3}{4}$ (c) $\frac{1}{2}$ (d) $\frac{1}{3}$
   (e) $\frac{2}{5}$ (f) $\frac{2}{3}$ (g) $\frac{5}{6}$ (h) $\frac{3}{5}$

(5) Compare fractions

Activity

Give your student a copy of the labeled fraction strips from the appendix. Or fold strips of paper into fourths and eights or draw bars.

Write the fractions $\frac{3}{4}$ and $\frac{5}{8}$ and ask your student which is greater.

Point out that we cannot compare them directly because neither the numerators nor the denominators are the same. We can see from the fraction bars that three fourths is larger than five-eighths. Ask your student how can we compare them without pictures. One way is to rename one or both fractions so that they have the same denominator. Looking at the denominators, we see that the denominator of five eighths is twice that of three fourths. We can rename three fourths by multiplying the numerator and denominator by 2 to get the equivalent fraction six eighths, which is greater than five eighths. So three fourths is the greater fraction. With the fraction bars, it is easy to see that we can just split each fourth part into two to get parts equal in size to eighths.

$\frac{3}{4}$ $\frac{5}{8}$

$\frac{3}{4}$

$\frac{5}{8}$

$\frac{3}{4} = \frac{6}{8}$ $\frac{5}{8} < \frac{6}{8}$

$\frac{5}{8} < \frac{3}{4}$

In summary, we can compare fractions by renaming one so that it has the same denominator as the other. Then, the size of the parts are the same, and we can simply compare the number of parts.

Repeat with other examples as needed, using examples where one denominator is a multiple of another.

Practice

Tasks 8-12, p. 95

8. $\frac{3}{4} = \frac{\mathbf{6}}{8}$. $\frac{\mathbf{3}}{\mathbf{4}}$ is greater
9. $\frac{2}{5} = \frac{\mathbf{4}}{10}$, $\frac{\mathbf{7}}{\mathbf{10}}$ is greater
10. (a) $\frac{5}{6}$ (b) $\frac{1}{2}$ (c) $\frac{7}{10}$
11. (a) $\frac{7}{10}$ (b) $\frac{5}{6}$ (c) $\frac{5}{9}$
12. (a) $\frac{1}{2}, \frac{5}{8}, \frac{3}{4}$ (b) $\frac{3}{10}, \frac{2}{5}, \frac{3}{5}$

Workbook

Exercise 8, p. 108 (Answers p. 107)

Enrichment

Ask your student to write some equivalent fractions for one half. You can write the denominator (use an even number) and have your student supply the numerator. If the fraction has a numerator that is half the denominator, it is equal to one half.

$\frac{1}{2}$ $\frac{2}{4}$ $\frac{3}{6}$ $\frac{4}{8}$ $\frac{5}{10}$ $\frac{6}{12}$

Now write the fraction $\frac{3}{8}$ and ask your student if it is less than or greater than one half. Since 3 is less than half of 8, it is less than one half. Write the fraction $\frac{5}{7}$ and ask her if that is greater than or less than one half. Half of 7 is more than half of 6 (3) and less than half of 8 (4). Since 5 is greater than 4, it must be greater than half of 7, so $\frac{5}{7}$ is greater than one half. Finally, ask your student to compare $\frac{3}{8}$ and $\frac{5}{7}$ by comparing each to a half.

$$\frac{3}{8} \qquad \frac{5}{7}$$

$$\frac{3}{8}<\frac{1}{2} \quad \frac{1}{2}<\frac{5}{7} \longrightarrow \frac{3}{8}<\frac{5}{7}$$

Have your student look at the fractions in tasks 10-11 and see if he can compare any of them by comparing both to one half.

Write the two fractions $\frac{1}{3}$ and $\frac{2}{5}$ and discuss ways to compare them without a diagram. Both are less than one half.

⇒ Since one numerator is twice the other, we could compare them by renaming one third as a fraction with a numerator of 2.

⇒ We can also list equivalent fractions for each until we get some with the same denominator. Guide your student in listing equivalent fractions by multiplying the numerator and denominator in order by 2, 3, 4, and 5. Then have your student look at each list for a fraction with the same denominator. We can use those to compare the two fractions.

You can point out that if he starts with the fraction with the greater denominator, he can list equivalent fractions until he gets one with a denominator that can be divided by the denominator of the other fraction. and then find an equivalent fraction with that denominator.

$$\frac{1}{3} \qquad \frac{2}{5}$$

$$\frac{1}{3}=\frac{2}{6} \quad \frac{2}{6}<\frac{2}{5} \longrightarrow \frac{1}{3}<\frac{2}{5}$$

$$\frac{2}{5}, \frac{4}{10}, \mathbf{\frac{6}{15}}, \frac{8}{20}, \frac{10}{25}$$

$$\frac{1}{3}, \frac{2}{6}, \frac{3}{9}, \frac{4}{12}, \mathbf{\frac{5}{15}}$$

$$\frac{5}{15}<\frac{6}{15} \longrightarrow \frac{1}{3}<\frac{2}{5}$$

⇒ You may also want to point out to your student that he can find equivalent fractions by multiplying the numerator and denominator of one fraction by the denominator of the other fraction. Multiply the numerator and denominator of $\frac{1}{3}$ by 5, which is the denominator of $\frac{2}{5}$, and multiply the numerator and denominator of $\frac{2}{5}$ by 3, which is the denominator of $\frac{1}{3}$, to rename both fractions into equivalent fractions with the same denominator, and compare those.

x 5   x 3

$$\frac{1}{3} = \frac{5}{15} \qquad \frac{2}{5} = \frac{6}{15}$$

x 5   x 3

$$\frac{5}{15}<\frac{6}{15} \longrightarrow \frac{1}{3}<\frac{2}{5}$$

Write the two fractions $\frac{5}{8}$ and $\frac{3}{5}$ and have your student use any of these strategies to compare them.

$$\frac{5}{8} \qquad \frac{3}{5}$$

$$\frac{5}{8}=\frac{25}{40} \quad \frac{3}{5}=\frac{24}{40} \longrightarrow \frac{3}{5}<\frac{5}{8}$$

(6) Practice

Practice

Practice B, p. 96

Unless you are allowing your student to write in the textbook, instruct her to write the greatest fraction, not circle it in problem 3.

3(c): In this problem, one denominator is not a simple multiple of the other. If your student cannot solve this, allow him to use fraction bars, or use this problem for the Enrichment activity below.

1. (a) $\frac{2}{8}$ (b) $\frac{9}{15}$ (c) $\frac{2}{6} = \frac{3}{9}$
   (d) $\frac{2}{5}$ (e) $\frac{2}{3}$ (f) $\frac{2}{4} = \frac{3}{6}$
2. (a) $\frac{4}{10}$ (b) $\frac{9}{12}$ (c) $\frac{4}{6} = \frac{6}{9}$
   (d) $\frac{1}{2}$ (e) $\frac{3}{4}$ (f) $\frac{3}{6} = \frac{5}{10}$
3. (a) $\frac{7}{10}$ (b) $\frac{5}{6}$ (c) $\frac{10}{12}$
   (a) $\frac{5}{6}$ (b) $\frac{2}{3}$ (c) $\frac{3}{4}$
4. (a) $\frac{1}{7}, \frac{3}{7}, \frac{5}{7}$ (b) $\frac{1}{10}, \frac{1}{5}, \frac{1}{2}$
   (c) $\frac{1}{2}, \frac{2}{3}, \frac{5}{6}$ (d) $\frac{1}{4}, \frac{5}{12}, \frac{2}{3}$
5. $\frac{1}{2}$ is greater than $\frac{2}{6}$.
   Sara ate a bigger portion.

Test

Tests, Unit 10, 2A and 2B, pp. 151-157

Reinforcement

Write 5-6 random numbers between 1 and 12 and have your student make as many fractions in their simplest form from them as possible. The numerator should always be smaller than the denominator. You can do this as a game. Use 4 sets of number cards 1-12. Shuffle and place face down in the middle. Each player draws an even number of cards, such as 6. The player who can use the most pairs to form simplest fractions wins the round.

12, 4, 5, 3, 8, 9

$\frac{4}{5}, \frac{4}{9}, \frac{5}{12}, \frac{5}{8}, \frac{5}{9}, \frac{3}{4}, \frac{3}{5}, \frac{3}{8}, \frac{8}{9}$

Enrichment

Write the two fractions $\frac{10}{12}$ and $\frac{4}{5}$. Ask your student to simplify $\frac{10}{12}$. Now we can compare $\frac{5}{6}$ and $\frac{4}{5}$. If you did the enrichment in the last lesson, he can rename each fraction to equivalent fractions that have the same denominator so that he can compare them directly.

Point out that the numerators for both are almost the same as the denominator. Both are only one part away from one. Ask her what fraction is needed to make 1 with each of them. Then ask her to compare those. $\frac{1}{6}$ is less than $\frac{1}{5}$ so $\frac{5}{6}$ is closer to 1 than $\frac{4}{5}$ is. $\frac{5}{6}$ must therefore be greater. You can use bars or a number line to illustrate this idea.

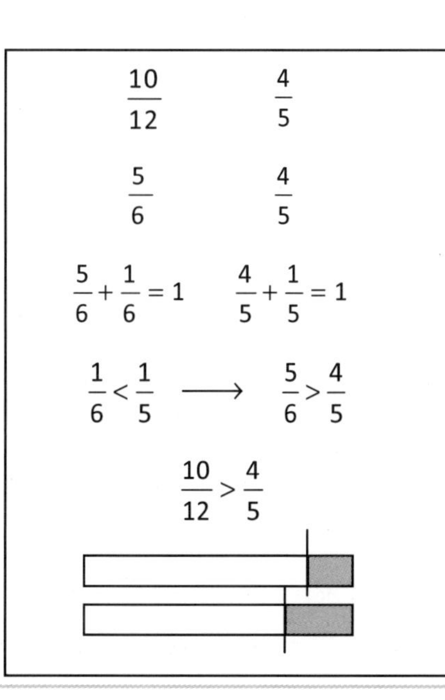

Chapter 3 – Adding Fractions

Objectives

- Add like fractions.
- Solve word problems involving addition of like fractions.

Vocabulary

- Like fractions
- Unlike fractions

Notes

In this chapter, students will be adding fractions with the same denominators. Since the denominators are the same, then the size of the parts are the same, and we add the fractions by simply adding the number of parts, the numerators.

In this curriculum, we will call fractions with the same denominator like fractions. Fractions where the denominators are different are called unlike fractions.

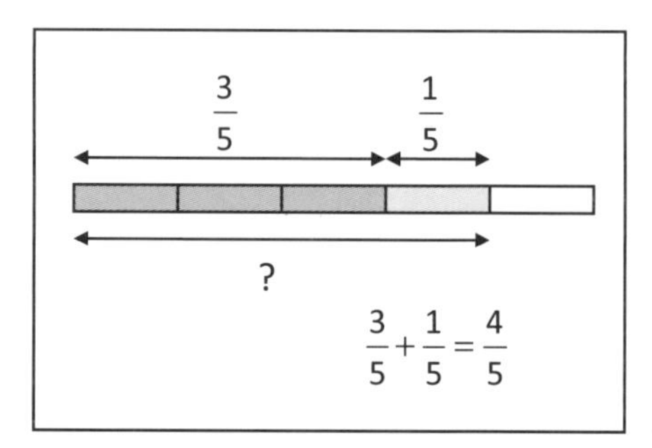

Since the fractions we are adding have the same denominator, and since at this level the sum of the numerators will never be more than 1, we can easily illustrate the process with fraction bars, which, not surprisingly, resemble bars students have drawn for division problems.

At this level, the entire bar represents a whole, and $\frac{1}{5}$ of the bar, for example, is simply $\frac{1}{5}$ of the whole. In *Primary Mathematics* 4, the bar will represent a set of objects, or a number, such as 20, and to show $\frac{1}{5}$ of 20 we divide the bar into 5 equal parts. Using fraction bars thus will lead nicely into using bar models to solve word problems involving fractions of a set and so are a better model in this curriculum than squares or circles. They also more closely resemble number lines, which you may also use to illustrate the addition and subtraction of fractions.

In *Primary Mathematics* 4A, students will learn to add related fractions, which are fractions where the denominator of one is a simple multiple of the other. To add related fractions, we just need to rename one of the fractions into an equivalent fraction with the same denominator as the other. In *Primary Mathematics* 5A, students will learn to add unrelated fractions, which are fractions where neither denominator is a simple multiple of the other. To add unrelated fractions, we need to rename both fractions into equivalent fractions where the denominators are the same.

If your student has trouble with the following lesson, illustrate more of the problems with fraction bars.

(1) Add like fractions

Discussion

Concept page 97

Read, or have your student read, the word problem. Ask your student how many fifths of liter are in one fifth of a liter and in two fifths of a liter, and then how many fifths of a liter there are altogether. Read the addition expression in words: "one fifth and two fifths is" and get your student to answer "three fifths." Saying it aloud can make it obvious that we are adding together fifths of a liter. All we have to do is add the number of fifths together.

$\frac{1}{5}\ \ell =$ ____ fifth of a liter

$\frac{2}{5}\ \ell =$ ____ fifths of a liter

$\frac{1}{5} + \frac{2}{5} = \frac{3}{5}$

Activity

Draw a bar and tell your student it represents one kilometer. Tell her that you walked for three eighths of a kilometer. Ask her how you would show that distance. Divide the bar into eighths and show the distance you walked. Then tell her after resting you walked another three eighths of a kilometer. Show that distance. Ask how far you walked in all. You walked six eights of a kilometer in all. Each distance you walked is three of the eighths, so altogether you walked six of the eighths. Write the equation. Ask her if you can say how far you walked more simply. Since six eighths is the same as three fourths, you can say you walked a total of three fourths of a kilometer.

1 km

$\frac{3}{8}$ km $\frac{3}{8}$ km

$\frac{3}{8} + \frac{3}{8} = \frac{6}{8} = \frac{3}{4}$

Tell your student that on a different day you walked four sevenths of a kilometer, rested, and then walked another two sevenths of a kilometer. Ask him how far you walked in all, and ask your student to write an addition equation. Then ask him which day you walked the farthest. Write the inequality shown here and have your student insert the correct symbol.

$\frac{4}{7} + \frac{2}{7} = \frac{6}{7}$

$\frac{3}{8} + \frac{3}{8}\ ?\ \frac{4}{7} + \frac{2}{7}$

$\frac{3}{8} + \frac{3}{8} < \frac{4}{7} + \frac{2}{7}$

Discussion

Tasks 1-2, pp. 97-98

1. $\frac{5}{5} = 1$

2. (a) $\frac{5}{8}$

   (b) $\frac{6}{8} = \frac{3}{4}$

Practice

Task 3, p. 98

3. (a) $\frac{5}{9}$ (b) $\frac{4}{7}$ (c) $\frac{5}{6}$

   (d) $\frac{2}{3}$ (e) 1 (f) $\frac{4}{5}$

   (g) 1 (h) $\frac{2}{3}$ (i) $\frac{1}{2}$

   (j) 1 (k) 1 (l) $\frac{2}{3}$

Workbook

Exercise 9, pp. 109-111 (Answers p. 107)

Reinforcement

Extra Practice, Unit 10, Exercise 3, pp. 185-186

Test

Tests, Unit 10, 3A and 3B, pp. 159-165

Chapter 4 – Subtracting Fractions

Objectives

- Subtract like fractions.
- Solve word problems involving subtraction of like fractions.

Notes

In this chapter, students will be subtracting fractions with the same denominators. Since the denominators are the same, the size of the parts are the same, we subtract the fractions by simply subtracting the number of parts, the numerators.

We can illustrate the process with bars. We use a bar to represent the whole, divide it into parts determined by the denominator, mark the number of parts that represent the whole, and the part we will subtract. The remaining part is the difference. Using fraction bars helps relate the process to the bar models students have been using for whole numbers, and also relates fractions to division, which will become more important as students do operations on fractions of a set.

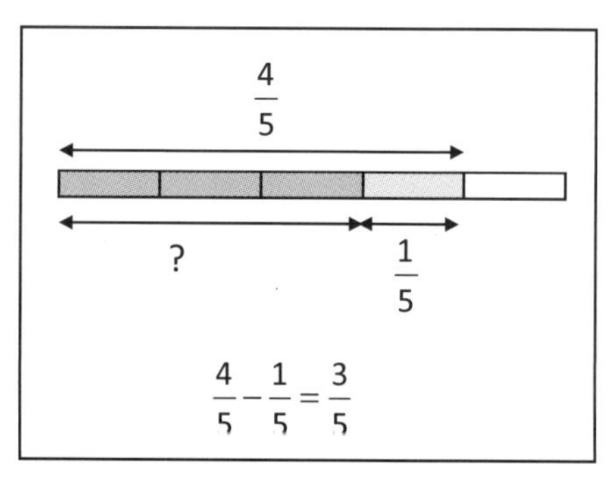

In *Primary Mathematics* 4A students will learn to subtract related fractions, and in *Primary Mathematics* 5A they will learn to subtract unrelated fractions.

If your student has trouble with the following lesson, illustrate more of the problems with fraction bars.

(1) Subtract like fractions

Activity

Draw a rectangle and divide it into twelfths, as shown here. Tell your student to imagine this is a pan of brownies or other food. Or use a real pan of brownies.

Ask your student how many parts there are and what fraction of the whole each part is. Remove or cross out 5 parts. Ask her what fraction was removed. Tell her that we started with a whole pan of brownies and ate or put on a plate $\frac{5}{12}$ of the pan. Ask her what fraction is left and then ask her to write an subtraction equation. In order to subtract five twelfths, we need to think of the whole as $\frac{12}{12}$. Then we can subtract the parts from the total number of parts.

Remove or cross out another 3 parts and ask your student to write another subtraction equation. Then ask her how we can write the remaining fraction more simply. Tell her that we always want to write the answers in simplest form.

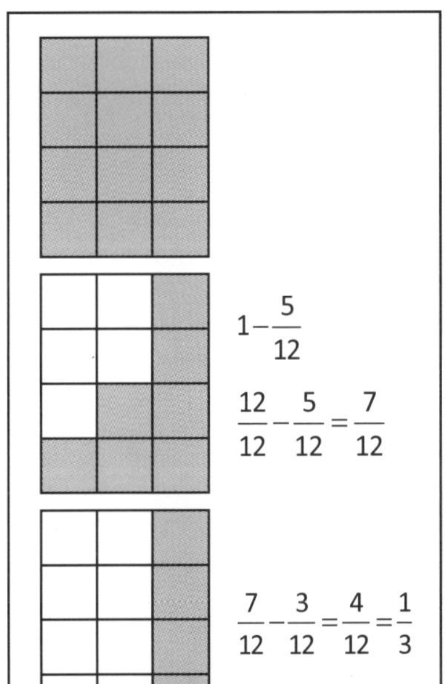

Discussion

Concept page 99

Tasks 1-3, pp. 99-100

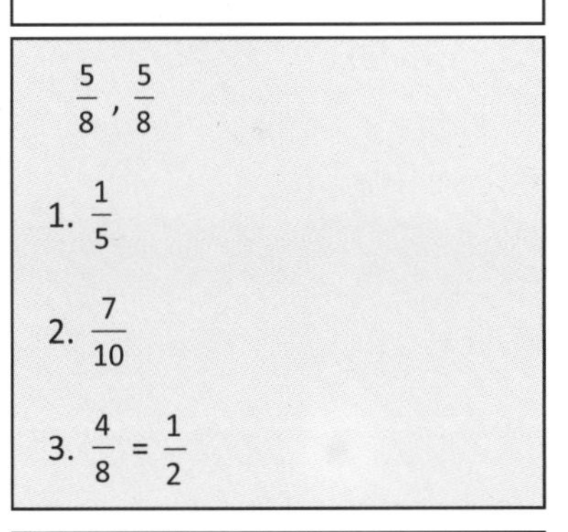

$\frac{5}{8}$, $\frac{5}{8}$

1. $\frac{1}{5}$

2. $\frac{7}{10}$

3. $\frac{4}{8}=\frac{1}{2}$

Activity

Write the problem shown here at the right and ask your student to fill in the correct symbol.

$1-\frac{2}{5}$? $\frac{5}{7}-\frac{3}{7}$

$1-\frac{2}{5} > \frac{5}{7}-\frac{2}{7}$

Practice

Task 4, p. 100

4. (a) $\frac{3}{5}$ (b) $\frac{1}{8}$ (c) $\frac{4}{9}$
(d) $\frac{1}{2}$ (e) $\frac{2}{5}$ (f) $\frac{1}{4}$
(g) $\frac{7}{9}$ (h) $\frac{1}{10}$ (i) $\frac{5}{12}$
(j) $\frac{1}{5}$ (k) $\frac{3}{8}$ (l) $\frac{1}{3}$

Workbook

Exercise 10, pp. 112-114 (Answers p. 108)

Reinforcement

Extra Practice, Unit 10, Exercise 4, pp. 187-188

(2) Practice

Practice

Practice C, p. 101

Point out that in all of the word problems, we are always given the fraction followed by the word *of*. A fraction is always a part of something. In problems 1-5, where we are not given a specific whole, we can assume any whole, and that all fractions in the same problem are of the same whole, and the answer is also a fraction of the same whole.

If you are having your student write answers encourage him to include the word *of* and what the whole is, even when he is not writing a full sentence for the answer.

Test

Tests, Unit 10, 4A and 4B, pp. 167-174

1. (a) $\frac{4}{5}$ (b) $\frac{5}{6}$ (c) $\frac{7}{10}$

2. (a) $\frac{3}{10}$ (b) $\frac{3}{7}$ (c) $\frac{7}{9}$

3. (a) $\frac{7}{8}$ (b) 1 (c) $\frac{7}{9}$

4. (a) $\frac{2}{5}$ (b) $\frac{2}{3}$ (c) $\frac{1}{2}$

5. (a) $\frac{3}{5}$ (b) $\frac{1}{2}$ (c) $\frac{8}{11}$

6. $\frac{1}{8}+\frac{3}{8}=\frac{4}{8}=\frac{1}{2}$

   They ate $\mathbf{\frac{1}{2}}$ **of the cake**.

7. $1-\frac{4}{9}=\frac{5}{9}$

   He saved $\mathbf{\frac{5}{9}}$ **of his money**.

8. $1-\frac{3}{7}=\frac{4}{7}$

   He spent $\mathbf{\frac{4}{7}}$ **of his money** on a racket.

9. $1-\frac{1}{6}-\frac{3}{6}=\frac{2}{6}=\frac{1}{3}$

   She had $\mathbf{\frac{1}{3}}$ **of the pie** left.

Chapter 5 – Fraction of a set

Objectives

- Find the number of objects in a fraction of a set.
- Solve simple word problems involving finding the fraction of a set.

Notes

In *Primary Mathematics* 2B, students were introduced to the concept of a fraction of a set, where the whole is a set of objects. The objects are divided into equal parts. Again, the numerator counts the number of parts, and the denominator is the total number of equal parts the whole has been divided into.

If there are 2 red counters and 8 blue counters, we can treat each counter as a part. The whole is 8 counters, and the fraction of counters that is red is then $\frac{2}{8}$ of the whole. We can also divide the set of 8 objects into 4 equal groups of 2. The two red counters consist of one group, so the fraction of counters that is red is $\frac{1}{4}$ of the whole. Note that $\frac{2}{8}$ of 8 and $\frac{1}{4}$ of 8 are the same. $\frac{2}{8}$ and $\frac{1}{4}$ are equivalent fractions, as long as the whole is the same. Fractions have no meaning without the whole. $\frac{1}{4}$ of 8 is not the same as $\frac{1}{4}$ of 12.

In *Primary Mathematics* 2B, students were given the amount in a part and the amount in the whole and asked to find what fraction of the whole the part was. In this chapter, they will be given the whole and the fraction of the whole, and asked to find the amount in the fraction.

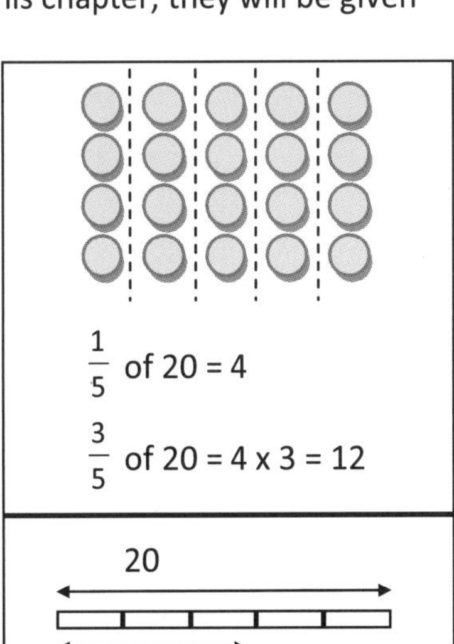

To find $\frac{1}{5}$ of a set of 20 objects, we can divide the set of 20 into 5 equal parts and determine how many objects there are in one part. To find $\frac{3}{5}$ of 20, we will first find $\frac{1}{5}$ of 20 and multiply that value by 3. Fractions express division. Your student will realize that we divide to find the fraction of a set. The equivalence of fractions and division will be formally taught in *Primary Mathematics* 4.

If your student struggles with using division to find the fraction of a set, allow her to use counters for the word problems.

If your student is having no problem with finding the fraction of a set, you can show him how to use a bar model. Draw a bar and label it with the whole, divide the bar into the number of parts, given by the denominator, and label the number of parts we need to find, given by the numerator, with a question mark.

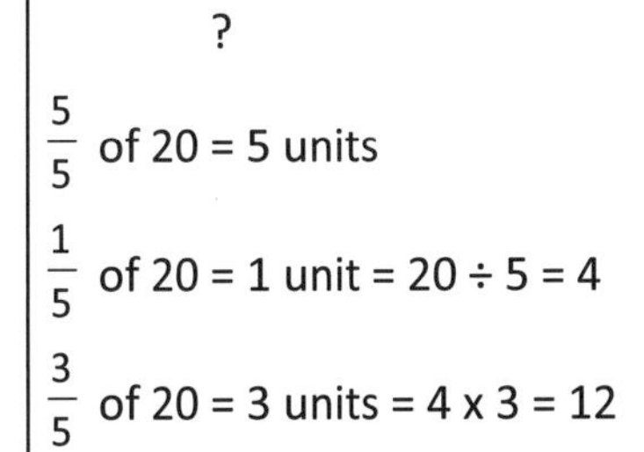

Material

- Counters

(1) Find the fraction of a set

Activity

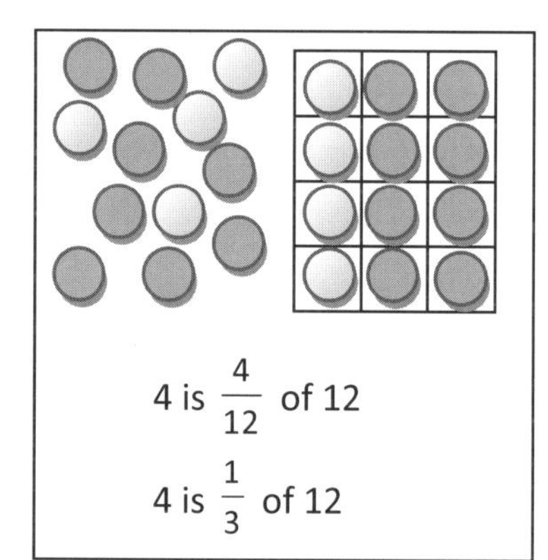

Give your student 4 counters of one color, such as yellow, and 8 of another color, such as blue. Ask him how many counters there are. Tell him that the 12 counters is the whole, just as if each counter stands for one brownie in a pan of brownies that have been cut into 12 equal pieces. You can arrange them as if they were a pan of brownies. Ask him what fraction of the counters are yellow. Since each counter is a part of the whole, and there are 12 equal parts, 4 out of 12 counters are green. So $\frac{4}{12}$ of the 12 counters are green.

Ask your student if we can write the fraction more simply. $\frac{4}{12}$ can be renamed as $\frac{1}{3}$. Ask her if 4 out of 12 is also $\frac{1}{3}$ of 12 and to explain her answer. We can put the counters into groups of 4, with all the yellow counters in one part. Then the yellow counters are one part out of 12, so $\frac{1}{3}$ of 12 is 4.

Discussion

Concept p. 102
Tasks 1-3, pp. 103-104

Write equations for these problems, as shown in the answers given here, which will lead into the concepts in the next lesson.

1(e): When discussing this problem, you can use actual counters. Make sure your student realizes that the circles are divided into parts in the picture, but the blue circles do not make up equal parts of that size. There are 8 out of 12 blue circles. Simplifying $\frac{8}{12}$ to $\frac{2}{3}$ tells us that if we divide the 12 into thirds, we could put the 8 blue circles into two of those thirds.

2: You can have your student use red and green counters to try to create equal groups where there are only red counters in some of the groups. For example, in 2(c), all the red apples could not go into a single group. 4 out of 6 are red, $\frac{4}{6}$ can be renamed as $\frac{2}{3}$, so we can make 3 groups and put the red apples in 2 of them.

$\frac{1}{2}$ of the animals are all white.

1. (a) $2 = \frac{2}{5}$ of 5
   (b) $1 = \frac{1}{3}$ of 3
   (c) $3 = \frac{1}{2}$ of 6
   (d) $2 = \frac{1}{5}$ of 10
   (e) $\frac{8}{12} = \frac{2}{3}$
   $8 = \frac{2}{3}$ of 12

2. (a) $5 = \frac{5}{6}$ of 6
   (b) $3 = \frac{1}{2}$ of 6
   (c) $2 = \frac{1}{3}$ of 6
   (d) $4 = \frac{2}{3}$ of 6

3. $\frac{2}{5}$, $\frac{2}{5}$

Workbook

Exercise 11, problems 1-2 only, p. 115 (Answers p. 108)

Reinforcement

Give your student a handful of counters of 3 or 4 different colors. Ask him to find what fraction of the total are each color, simplifying where possible and arranging in groups to verify.

(2) Find the value of a fraction of a set

Activity

Give your student 24 counters all of the same color. Ask her to find out how many counters are in $\frac{1}{6}$ of 24. $\frac{1}{6}$ is the number of counters in 1 part out of 6 equal parts. So we need to make 6 equal parts. Get your student to divide the counters into 6 equal groups and determine the number in each group. There are 4 counters in one part. So $\frac{1}{6}$ of 24 = 4. Point out that finding $\frac{1}{6}$ of 24 is the same as dividing 24 by 6.

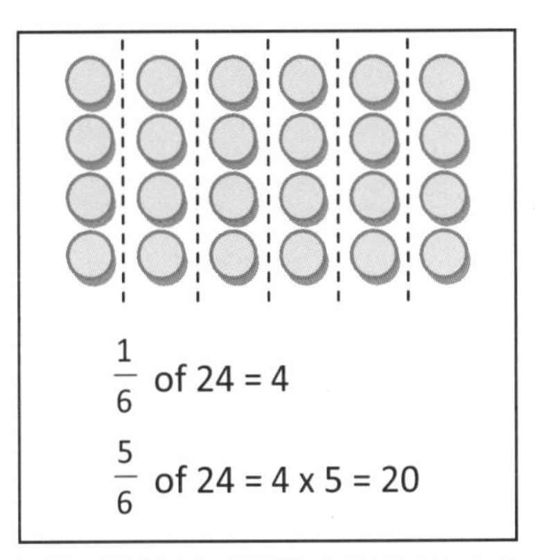

$\frac{1}{6}$ of 24 = 4

$\frac{5}{6}$ of 24 = 4 x 5 = 20

Now ask your student to find out how many counters are in $\frac{5}{6}$ of 24. Again, the 6 in the denominator means we have 6 equal parts. The 5 in the numerator means 5 of those parts. Since one part has 4 counters, then 5 parts has 4 x 5 counters.

Ask your student to use the counters to find $\frac{3}{4}$ of 24, $\frac{3}{8}$ of 24, and $\frac{5}{12}$ of 24.

$\frac{1}{4}$ of 24 = 24 ÷ 4 = 6

$\frac{3}{4}$ of 24 = 6 x 3 = 18

$\frac{1}{8}$ of 24 = 24 ÷ 8 = 3

$\frac{3}{8}$ of 24 = 3 x 3 = 9

$\frac{1}{12}$ of 24 = 24 ÷ 12 = 2

$\frac{5}{12}$ of 24 = 2 x 5 = 10

Discussion

Tasks 4-5, p. 104

4: Ask your student to also find two thirds of 9.

4. $\frac{1}{3}$ of 9 = 3; $\frac{2}{3}$ of 9 = 6

5. 4; 12

Workbook

Exercise 11, problem 3, p. 116 (Answers p. 108)

Reinforcement

Extra Practice, Unit 10, Exercise 5, pp. 189-190

Test

Tests, Unit 10, 5A and 5B, pp. 175-182

Enrichment

Tell your student that we can use bar models to represent finding the fraction of a whole. Illustrate this with $\frac{3}{5}$ of 20. The whole is the entire bar. The number of equal units is the number in the denominator of the fraction. The number of units we want to find is the number in the numerator of the fraction.

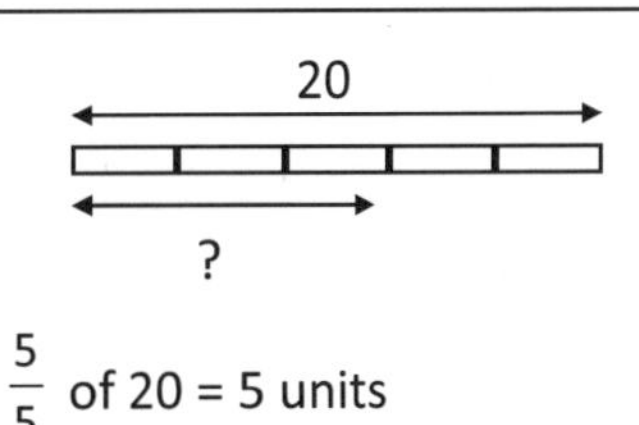

$\frac{5}{5}$ of 20 = 5 units

$\frac{1}{5}$ of 20 = 1 unit = 20 ÷ 5 = 4

$\frac{3}{5}$ of 20 = 3 unit = 4 x 3 = 12

Have your student draw bar models to find other fractions of a whole. Only use problems where the denominator of the fraction divides evenly into the whole.

Chapter 6 – Fractions and Money

Objectives

- Express a set of coins as a fraction of a dollar.
- Write an amount of money less than $1 using either the decimal notation or as a fraction of $1.
- Simplify fractions with 100 in the denominator.
- Using a set of coins as a set of objects, express the number of one type of coin as a fraction of the total coins.

Notes

In this chapter, students will write amounts of money less than a dollar as a fraction of a dollar.

Students have learned how many pennies, nickels, dimes, and quarters are in a dollar. In a set of coins of the same denomination where the total is a dollar, each coin is an equal part, so a given number of coins out of the number of that type of coin that makes up a dollar can be expressed as a fraction of a dollar. For example, since there are 20 nickels in a dollar, then three nickels is $\frac{3}{20}$ of a dollar. The dollar is the whole.

A set of coins of different denominations can be expressed as fraction of a dollar by putting the number of cents over 100, since there are 100 cents in a dollar, and simplifying, or by renaming the coins into a coin of a different denomination. For example, 2 dimes and 1 nickel is the same as 25 cents, and so is $\frac{25}{100}$ or $\frac{1}{4}$ of a dollar. 2 dimes and a nickel is also the same as a quarter. Since there are 4 quarters in a dollar, one quarter, (or 2 dimes and 1 nickel) is $\frac{1}{4}$ *of a dollar.*

Students will also be asked to look at the set of coins as simply a set of different objects, rather than money. If we have 2 dimes and 1 nickel, then $\frac{2}{3}$ *of the coins* are dimes. The student must therefore distinguish between the dollar as the whole and finding the amount in cents as a fraction of the dollar, or the number of coins as the whole and finding the type of coin as a fraction of a set of coins.

In working with cents as a fraction of a dollar, we are working with a set of equivalent fractions with denominators of 100, 50, 25, 20, 10, 5, 4, and 2.

This chapter brings together several important concepts: counting coins, changing money to different denominations, writing the amount of cents in dollars using a dot (decimal), grouping into equal parts to find the fraction of a whole, simplifying fractions, and the importance of what the whole is for a fraction. It also provides a concrete introduction to the relationship between fractions and decimals. This is just an introduction; do not try to explain what decimal numbers are at this point. Decimal numbers and the relationship between fractions and decimals will be formally taught in *Primary Mathematics* 4B.

Material

- Coins (4 quarters, 20 nickels, 10 dimes, 100 pennies)
- Counters, if you don't have 100 pennies available

(1) Find the fraction of a dollar for a set of one kind of coin

Discussion

Concept pages 105-106

10 pennies = $\frac{10}{100}$ or $\frac{1}{10}$ of a dollar = \$0.10

30 pennies = $\frac{30}{100}$ or $\frac{3}{10}$ of a dollar = \$0.30

Use 100 actual pennies, or counters to stand for pennies, when discussing this page. With actual objects, it is easier to create equal groups than just looking at the picture in the book and imagining the groups.

Since there are 100 pennies in a dollar, 1 penny is 1 cent out of 100 cents in the whole, or $\frac{1}{100}$ of a dollar. Since we can also represent 1 penny or 1 cent as \$0.01, $\frac{1}{100}$ *of a dollar* and \$0.01 both mean the same amount, 1 cent.

To find 10 pennies as a fraction of a dollar, we can group the 100 pennies by tens to have 10 parts, with 10 pennies in one part, or rename $\frac{10}{100}$ as $\frac{1}{10}$. 10 pennies, or 10 cents, is $\frac{1}{10}$ *of a dollar*. 30 pennies is then 3 parts out of 10, or $\frac{3}{10}$ of a dollar.

Get your student to also write 10 cents and 30 cents in dollars (decimal notation).

1 dime is the same as 10 pennies, so each dime is a part with 10 cents in it, and it is the same fraction of a dollar as 10 pennies. Likewise, 3 dimes is the same fraction of a dollar as 30 pennies; both are 30 cents.

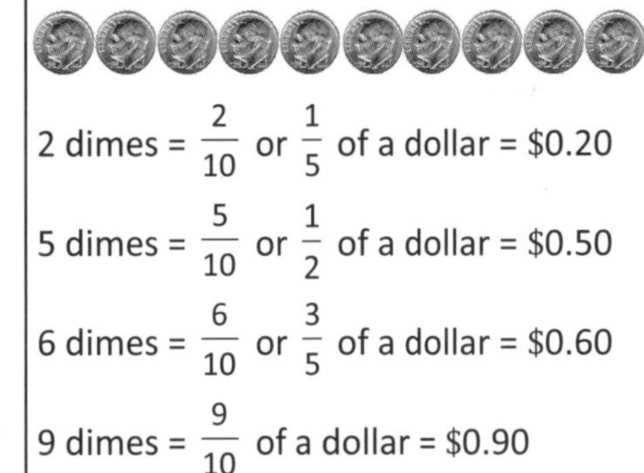

2 dimes = $\frac{2}{10}$ or $\frac{1}{5}$ of a dollar = \$0.20

5 dimes = $\frac{5}{10}$ or $\frac{1}{2}$ of a dollar = \$0.50

6 dimes = $\frac{6}{10}$ or $\frac{3}{5}$ of a dollar = \$0.60

9 dimes = $\frac{9}{10}$ of a dollar = \$0.90

Activity

Use the 10 dimes and ask your student to find what fraction of a dollar 2, 5, 6, and 9 dimes are. He should express the answer in simplest form. He can find the answer by simplifying the fraction out of 10 when possible, or by making equal groups, as was done in the first lesson in the last chapter. Have him write the amount both as a fraction of a dollar and in dollars.

Ask your student how many nickels there are in a dollar. Provide her with 20 nickels. Ask her what fraction of a dollar 1 nickel is. It is 1 out of the 20 nickels in a dollar, or $\frac{1}{20}$ of a dollar. 5 cents is $\frac{1}{20}$ of a dollar. Ask her what fraction of a dollar is 2 nickels. 2 nickels is 2 out of the 20 equal parts in a dollar, or 1 out of 10 equal parts if we put two nickels in each part. Tell her that since 2 nickels is the same as a dime, we can

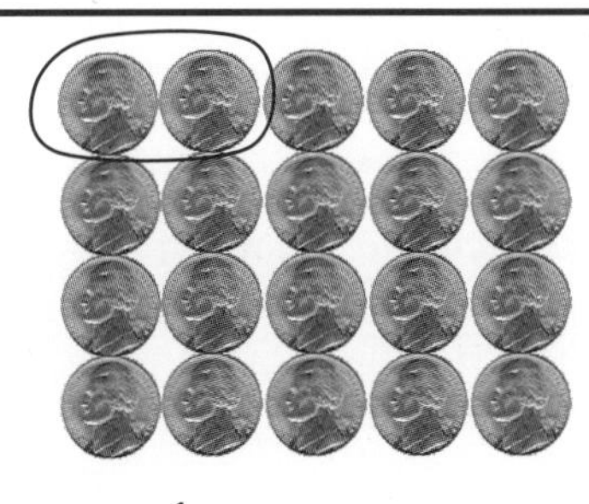

1 nickel = $\frac{1}{20}$ of a dollar = \$0.05

2 nickels = $\frac{2}{20}$ or $\frac{1}{10}$ of a dollar = \$0.10

easily find what fraction of a dollar 2 nickels is by renaming it as 1 dime.

Ask your student to find what fraction of a dollar 3, 6, and 10, nickels are, and to write the amount in dollars. She can write the fractions as number of nickels out of 20, or, if the number of nickels is even, convert to dimes and write the fraction as number of dimes out of 10, and then simplify.

3 nickels = $\frac{3}{20}$ of a dollar = \$0.15

6 nickels = $\frac{6}{20}$ or $\frac{3}{10}$ of a dollar = \$0.30

10 nickels = $\frac{10}{20}$ or $\frac{1}{2}$ of a dollar = \$0.50

Discussion

Tasks 1-3, pp. 106-107

1. \$1; \$0.25 2. \$0.50 3. \$0.75

Ask your student how many quarters are in a dollar, and then what fraction of a dollar 1 quarter is. 1 quarter is one out of 4 quarters in a dollar, or one fourth of a dollar. Ask her to write the amount in dollars. Continue with 2 quarters and 3 quarters.

1 quarter = $\frac{1}{4}$ of a dollar = \$0.25

2 quarters = $\frac{2}{4}$ or $\frac{1}{2}$ of a dollar = \$0.50

3 quarters = $\frac{3}{4}$ of a dollar = \$0.75

Activity

Ask your student what fraction of a dollar 5 nickels is. Point out that 5 nickels is the same as a quarter. We can put the 20 nickels into 4 equal groups, with 5 nickels in one groups. So 5 nickels is the same fraction of a dollar as 1 quarter.

5 nickels = 1 quarter = $\frac{1}{4}$ of a dollar

Provide your student with 100 pennies. Ask him to find what fraction of a dollar 5 pennies, 15 pennies, 20 pennies, 25 pennies, and 50 pennies are, and also write the equivalent amount in dollars. He can convert the pennies mentally or physically into nickels, dimes, or quarters when possible, and find the fraction that way. Or he can explore to determine the best way to form groups in order to get equal parts, which should be fairly obvious what groups are needed from having worked with the other denominations.

5 pennies = $\frac{5}{100}$ or $\frac{1}{20}$ of a dollar = \$0.05

15 pennies = $\frac{15}{100}$ or $\frac{3}{20}$ of a dollar = \$0.15

20 pennies = $\frac{20}{100}$ or $\frac{1}{5}$ of a dollar = \$0.20

25 pennies = $\frac{25}{100}$ or $\frac{1}{4}$ of a dollar = \$0.25

50 pennies = $\frac{50}{100}$ or $\frac{1}{2}$ of a dollar = \$0.50

Enrichment

Ask your student to find what fraction of a dollar 32 pennies is in simplest form. In order to simplify 32 out of 100, she needs to try to group 100 pennies into groups such that there are 32 pennies in some number of the groups. Since 32 is an even number, you can suggest grouping by 2's, then combining groups to group by 4's. There are 25 groups of 4 now, or 25 parts. She cannot combine the groups of 4 into 8's and end up with equal groups. 32 pennies are in 8 of the 25 parts. So the simplest fraction for $\frac{32}{100}$ is $\frac{8}{25}$. With 7 pennies, there is no way to make groups and have the 7 pennies be in equal parts.

32 pennies = $\frac{32}{100}$ or $\frac{8}{25}$ of a dollar = \$0.32

7 pennies = $\frac{7}{100}$ of a dollar = \$0.07

(2) Find the fraction of a dollar for a set of coins less than $1

Discussion

Tasks 4-8, p. 107

7: Your student can realize that 2 dimes and 1 nickel are same as a quarter, which is a fourth of a dollar since there are 4 quarters in a dollar, or he can think in terms of cents and simplify 25 out of 100 cents.

4. $\frac{1}{4}$ of a dollar

5. \$0.70 = $\frac{7}{10}$ of a dollar

6. $\frac{1}{2}$ of a dollar

7. $\frac{1}{2}$ of a dollar

8. $\frac{1}{4}$ of a dollar

Activity

Write the following amounts and ask your student what fraction of a dollar each amount is.

⇒ \$0.25: Your student could remember this is a quarter, or think of it as 25 cents.

⇒ \$0.75: This is 3 quarters.

⇒ \$0.15: Encourage your student to think of this as 3 nickels, so it is three twentieths of a dollar. Or he can think of it as 15 cents and simplify the fraction fifteen hundredths by dividing the numerator and denominator by 5.

⇒ \$0.09: This is 9 cents out of 100 and cannot be simplified.

\$0.25 = $\frac{25}{100}$ or $\frac{1}{4}$ of a dollar

\$0.75 = $\frac{75}{100}$ or $\frac{3}{4}$ of a dollar

\$0.15 = $\frac{15}{100}$ or $\frac{3}{20}$ of a dollar

\$0.09 = $\frac{9}{100}$ of a dollar

Ask your student how much money the following fractions are.

⇒ $\frac{1}{4}$ of a dollar. Recognize this as a quarter.

⇒ $\frac{3}{10}$ of a dollar. $\frac{1}{10}$ of a dollar is a dime, so $\frac{3}{10}$ is 3 dimes.

⇒ $\frac{2}{5}$ of a dollar. Rename the fraction as tenths; this is the same as 4 dimes.

$\frac{1}{4}$ of a dollar = \$0.25

$\frac{3}{10}$ of a dollar = \$0.30

$\frac{2}{5}$ of a dollar = $\frac{4}{10}$ of a dollar = \$0.40

Give your student 4 dimes and 2 nickels. Ask her what fraction of a dollar this is. Then ask her what fraction of the total coins are dimes. Make sure your student understands that now we are not finding the fraction *of a dollar*. Each coin is now a part, and its value is not important. It is as if we had 4 red counters and 2 blue counters and are told to find what fraction of the total counters is red. So now the whole is 6, and four sixths *of the coins* are dimes.

50 cents = $\frac{1}{2}$ of a dollar

$\frac{4}{6}$ or $\frac{2}{3}$ of the coins are dimes.

$\frac{1}{3}$ of the coins are nickels

Workbook

Exercise 12, p. 117 (Answers p. 109)

Reinforcement

Extra Practice, Unit 10, Exercise 6, pp. 191-192

(3) Practice

Practice

Practice D, p. 108

Reinforcement

Give your student some quarters, nickels, and dimes where the total amount of money is less than a dollar and ask him to find the amount as a fraction of a dollar. You can also ask what fraction of the coins are dimes, or nickels, or quarters, depending on the coins.

Test

Tests, Unit 10, 6A and 6B, pp. 183-188

Enrichment

Ask your student what numbers 100 can be divided by. 100 can be divided by 2, 5, 10, 20, 25, or 50. So we can only simplify a fraction with 100 in the denominator if the numerator can be divided by one of these numbers. The numerator has to be even, or end in 5. Discuss ways to simplify fractions with 100 in the denominator.

1) From the previous lesson, if the numerator is a ten, or is 25 or 75, it is easy to simplify by thinking in terms of dimes or quarters.
2) If the numerator ends in 5, but is not 25 or 75, we can divide by 5, and the denominator is 20. Let your student experiment with fractions with 15, 35, 45, 55, 65, 85, and 95 in the numerator to see that these cannot be further simplified. If we had those amounts in nickels, we could not trade them in for all dimes or quarters.
3) If the numerator is even, we can start by dividing both the numerator and denominator by 2 and then see if we can simplify another step.
4) If the numerator is not even or does not end in 5, the fraction (with 100 in the denominator) cannot be simplified.

1. (a) 24 coins

   (b) $\frac{1}{6}$ of the coins are quarters (1 part out of 6)

   (c) $\frac{1}{2}$ of the coins are dimes (3 parts out of 6)

   (c) $\frac{1}{3}$ of the coins are nickels (2 parts out of 6)

2. (a) $\frac{1}{4}$ of a dollar (1 quarter out of 4)

   (b) $\frac{1}{2}$ of a dollar (2 quarters out of 4, or 1 half-dollar out of 2)

   (c) $\frac{1}{10}$ of a dollar (1 dime out of 10)

   (d) $\frac{3}{4}$ of a dollar (3 quarters out of 4)

   (e) $\frac{3}{10}$ of a dollar (3 dimes out of 10)

   (f) 1 (1 whole dollar)

3. $\frac{8}{20}$ of a dollar = $\frac{2}{5}$ of a dollar

4. (a) $\frac{1}{7}$ of the coins are quarters

   (b) $\frac{3}{7}$ of the coins are dimes

   (c) $\frac{3}{7}$ of the coins are nickels

   (d) $0.70

   (e) $\frac{7}{10}$ of a dollar

1) $\frac{75}{100} = \frac{3}{4}$ $\quad$ $\frac{60}{100} = \frac{6}{10} = \frac{3}{5}$

2) $\frac{65}{100} = \frac{13}{20}$ $\quad$ $\frac{85}{100} = \frac{17}{20}$

3) $\frac{32}{100} = \frac{16}{50} = \frac{8}{25}$ $\quad$ $\frac{48}{100} = \frac{24}{50} = \frac{12}{25}$

4) $\frac{11}{100}$ $\quad$ $\frac{63}{100}$

Review 10

Review

Review 10, pp. 109-111

Workbook

Review 10, pp. 118-122 (Answers p. 109)

Test

Tests, Units 1-10, Cumulative A and B, pp. 189-199

1. (a) 9210
   (b) 4060
2. (a) six thousand, two hundred four
   (b) three thousand, five hundred forty
   (c) five thousand twenty-eight
3. 3900
4. (a) 4014, 4041, 4104, 4410
   (b) 1112, 2111, 2121, 2211
5. 125 x 8 = **1000**
6. 500 ÷ 8 = 62 r4
7. 55 – ? = 44 ↔ 55 – 44 = **11**
8. (a) 20 (10 $10 in $100, so twice that in $200)
   (b) 30 (20 in $1, plus half that in 50 cents)
9. (a) $\frac{\mathbf{3}}{12}$ (b) $\frac{6}{\mathbf{9}}$ (c) $\frac{4}{\mathbf{5}}$
10. (a) $\frac{1}{4}$ (b) $\frac{2}{7}$ (c) $\frac{11}{12}$
    (d) $\frac{5}{8}$ (e) $\frac{3}{10}$ (f) $\frac{3}{6}$

11. (a) 420 cm (b) 2 m 5 cm
    (c) 2095 m (d) 1 km 600 m
    (e) 1040 g (f) 2 kg 450 g
    (g) 3060 ml (h) 2 ℓ 525 ml
    (i) 32 ft (j) 1 ft 5 in.
    (k) 138 oz (l) 1 lb 2 oz
    (m) 25 pt (n) 7 pt 1 c
12. Cost of 1 pound: $0.70
    Cost of 6 pounds: $0.70 x 6 = $4.20
    Change: $10 – $4.20 = $5.80
    He received **$5.80** in change.
13. $38.40 – $9.60 = $28.80
    The tennis racket costs $**28.80** more.
14. 1 ℓ – 325 ml = 675 ml
    There was **675 ml** of yogurt left.
15. 10 x 125 ml = 1250 ml = 1 ℓ 250 ml
    The total amount of milk is **1 ℓ 250 ml**.
16. $1 - \frac{4}{9} = \frac{5}{9}$
    He saved $\frac{\mathbf{5}}{\mathbf{9}}$ of his allowance.
17. $1 - \frac{3}{7} = \frac{4}{7}$
    He spent $\frac{\mathbf{4}}{\mathbf{7}}$ of his money on the racket.
18. (a) She donates **25¢** of each dollar.
    (b) $\frac{1}{4}$ of $8 = $2.
    She donates **$2**.
19. (a) yes
    (b) yes
    (c) no
    (d) yes (it is least likely to get a red ball)
    (e) yes
    (f) no

Workbook

Exercise 1, pp. 90-95

1. (a) **3** out of the **4** equal parts
   (b) **4** quarters
   3 quarters
   (c) $\frac{1}{4}$

2. (a) **4** out of the **6** equal parts
   (b) **6** sixths
   4 sixths
   (c) $\frac{2}{6}$

3. (a) **3** out of the **10** equal parts
   (b) **10** tenths
   3 tenths
   (c) $\frac{7}{10}$

4. (a) $\frac{1}{5}$
   (b) $\frac{4}{9}$

5. $\frac{4}{7}$ $\frac{5}{7}$ $\frac{7}{8}$ $\frac{3}{8}$ $\frac{9}{10}$ $\frac{7}{10}$

   $\frac{2}{7}$ $\frac{3}{10}$ $\frac{3}{7}$ $\frac{5}{8}$ $\frac{1}{8}$ $\frac{1}{10}$

6. (a) $\frac{2}{3}$
   (b) $\frac{3}{4}$
   (c) $\frac{3}{5}$
   (d) $\frac{4}{6}$
   (e) $\frac{5}{8}$
   (f) $\frac{5}{9}$

7. (a) **4** quarters = $\frac{\mathbf{4}}{4}$
   2 quarters
   (b) **5** fifths = $\frac{\mathbf{5}}{5}$
   4 fifths
   (c) **6** sixths = $\frac{\mathbf{6}}{6}$
   3 sixths
   (d) **8** eighths = $\frac{\mathbf{8}}{8}$
   7 eighths
   (e) **10** tenths = $\frac{\mathbf{10}}{10}$
   6 tenths
   (f) **12** twelfths = $\frac{\mathbf{12}}{12}$
   9 twelfths

8.

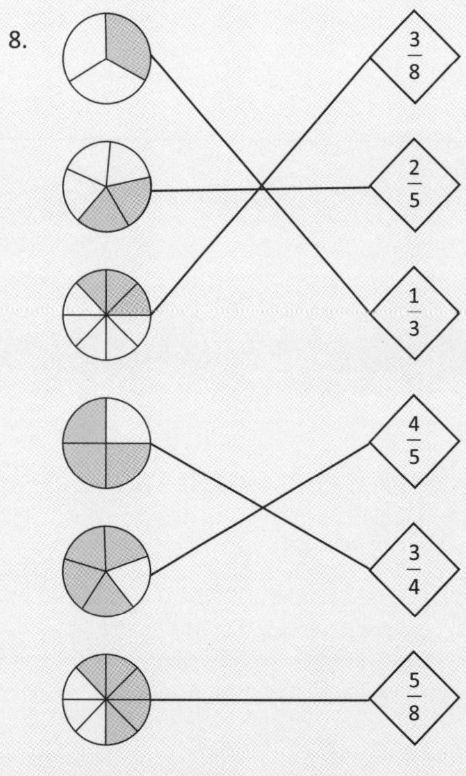

9. (a) $\frac{2}{3}$ (b) $\frac{7}{8}$
   (c) $\frac{5}{9}$ (d) $\frac{7}{10}$

10. Check how many parts are colored.

Workbook

Exercise 2, pp. 96-97

1. (a) $\frac{1}{3}$ (b) $\frac{1}{6}$
   (c) $\frac{3}{4}$ (d) $\frac{2}{3}$

2. (a) $\frac{1}{6}$ (b) $\frac{1}{5}$
   (c) $\frac{5}{8}$ (d) $\frac{3}{10}$

3. (a) $\frac{1}{7}$ (b) $\frac{1}{8}$
   (c) $\frac{6}{7}$ (d) $\frac{7}{8}$

4. (a) $\frac{1}{5}$ (b) $\frac{1}{10}$
   (c) $\frac{3}{7}$ (d) $\frac{5}{12}$

5. (a) $\frac{1}{10}, \frac{1}{7}, \frac{1}{6}$
   (b) $\frac{3}{10}, \frac{3}{8}, \frac{3}{4}$
   (c) $\frac{1}{9}, \frac{1}{5}, 1$

6. (a) $\frac{1}{3}, \frac{1}{4}, \frac{1}{12}$
   (b) $\frac{5}{7}, \frac{5}{9}, \frac{5}{12}$
   (c) $\frac{1}{8}, \frac{1}{10}, 0$

Exercise 3, pp,. 98-99

1. (a) $\frac{4}{6}$ $\frac{5}{6}$
   (b) $\frac{5}{8}$ $\frac{6}{8}$
   (c) $\frac{8}{12}$ $\frac{10}{12}$ 1
   (d) $\frac{6}{9}$ $\frac{5}{9}$ $\frac{4}{9}$
   (e) $\frac{6}{10}$ $\frac{5}{10}$ $\frac{4}{10}$

2. (a) $\frac{5}{8}$ (b) $\frac{2}{6}$
   (c) $\frac{4}{5}$ (d) $\frac{7}{10}$

3. (a) $\frac{4}{5}$ (b) $\frac{6}{7}$
   (c) $\frac{7}{10}$ (d) $\frac{5}{6}$

4. (a) $\frac{1}{3}$ (b) $\frac{1}{5}$
   (c) $\frac{4}{10}$ (d) $\frac{5}{12}$

5. (a) $\frac{4}{5}$ (b) $\frac{6}{7}$
   (c) $\frac{8}{9}$ (d) $\frac{10}{12}$

6. (a) $\frac{1}{4}$ (b) $\frac{2}{6}$
   (c) $\frac{4}{10}$ (d) $\frac{2}{11}$

7. (a) $\frac{3}{10}, \frac{5}{10}, \frac{8}{10}$
   (b) $\frac{3}{12}, \frac{5}{12}, 1$

Workbook

Exercise 4, pp. 100-101

1.

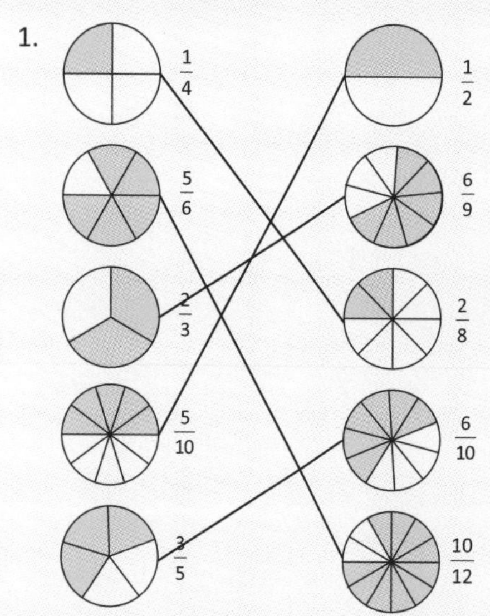

2. (a) $\frac{\mathbf{2}}{4}$ (b) $\frac{\mathbf{3}}{6}$ (c) $\frac{\mathbf{5}}{10}$

   (d) $\frac{\mathbf{2}}{6}$ (e) $\frac{\mathbf{4}}{6}$ (f) $\frac{\mathbf{10}}{10}$

   (g) $\frac{\mathbf{2}}{8}$ (h) $\frac{\mathbf{4}}{8}$ (i) $\frac{\mathbf{6}}{8}$

   (j) $\frac{\mathbf{2}}{10}$ (k) $\frac{\mathbf{4}}{10}$ (l) $\frac{\mathbf{8}}{10}$

Exercise 5, pp. 102-103

1. (a) $\frac{1}{2} = \frac{2}{\mathbf{6}} = \frac{\mathbf{3}}{9}$

   (b) $\frac{3}{4} = \frac{\mathbf{6}}{8} = \frac{9}{\mathbf{12}}$

   (c) $1 = \frac{\mathbf{3}}{3} = \frac{6}{\mathbf{6}}$

2. (a) $\frac{8}{10}$

   (b) $\frac{4}{12}$

3.

$\frac{1}{2}$	$\frac{1}{5}$	$\frac{9}{10}$	$\frac{4}{5}$	$\frac{8}{10}$
$\frac{2}{4}$	$\frac{2}{12}$	$\frac{2}{10}$	$\frac{4}{10}$	$\frac{1}{4}$
$\frac{2}{6}$	$\frac{1}{3}$	$\frac{2}{3}$	$\frac{3}{8}$	$\frac{2}{8}$
$\frac{1}{6}$	$\frac{1}{2}$	$\frac{1}{12}$	$\frac{2}{6}$	$\frac{5}{9}$
$\frac{5}{10}$	$\frac{5}{6}$	$\frac{2}{9}$	$\frac{3}{9}$	$\frac{4}{9}$
$\frac{11}{12}$	$\frac{3}{5}$	$\frac{6}{10}$	$\frac{6}{7}$	$\frac{5}{7}$

Exercise 6, pp. 104-105

1. (a) $\frac{4}{5}$ (b) $\frac{5}{6}$

   (c) $\frac{3}{4}$ (d) $\frac{8}{10}$

   (e) $\frac{1}{2}$ (f) $\frac{2}{3}$

   (g) $\frac{6}{12}$ (h) $\frac{2}{3}$

2. (a) $\frac{4}{6}$

   (b) $\frac{8}{10}$

   (c) $\frac{2}{5}$

   (d) $\frac{6}{6}$

   (e) $\frac{3}{4}$

   (f) $\frac{2}{12}$

   (g) $\frac{3}{4}$

   (h) $\frac{5}{10}$

Workbook

Exercise 7, pp. 106-107

1. (a) $\frac{2}{3}$ (b) $\frac{3}{4}$

2. (a) $\frac{1}{2}$ (b) $\frac{2}{3}$ (c) $\frac{1}{3}$

3. LUCKY

4.

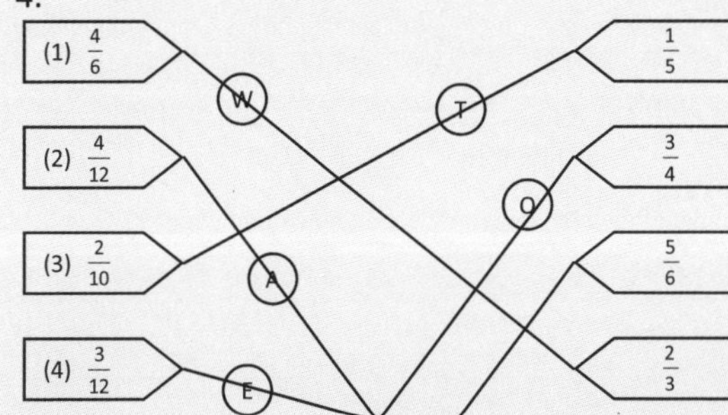
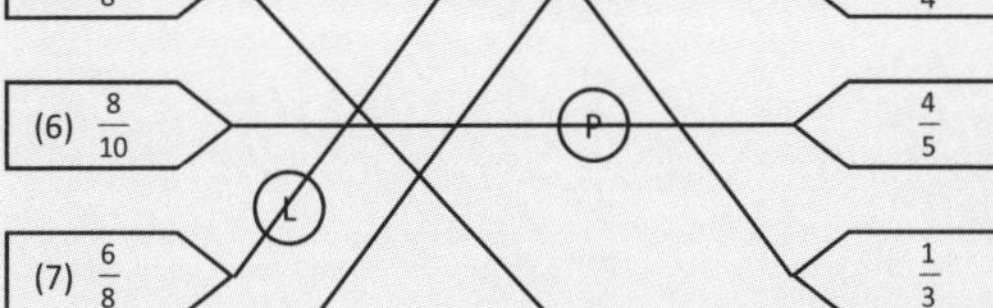
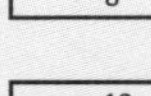

W	A	T	E	R	P	O	L	O
1	2	3	4	5	6	7	8	9

Exercise 8, p. 108

1. (a) $\frac{7}{8}$ (b) $\frac{4}{5}$
   (c) $\frac{2}{3}$ (d) $\frac{2}{3}$
   (e) $\frac{4}{5}$ (f) $\frac{11}{12}$
   (g) $\frac{2}{3}$ (h) $\frac{1}{2}$

2. (a) $\frac{1}{2}, \frac{2}{3}, \frac{5}{6}$
   (b) $\frac{1}{2}, \frac{5}{8}, \frac{3}{4}$
   (c) $\frac{7}{12}, \frac{2}{3}, \frac{5}{6}$
   (d) $\frac{7}{12}, \frac{2}{3}, \frac{3}{4}$

Exercise 9, pp. 109-111

1. (a) $\frac{3}{5}$ (b) $\frac{7}{8}$
   (c) $\frac{5}{6}$ (d) $\frac{7}{10}$

2. (a) 1 (b) $\frac{1}{2}$ (c) $\frac{2}{3}$
   (d) $\frac{3}{5}$ (e) $\frac{5}{6}$ (f) $\frac{5}{7}$
   (g) $\frac{3}{4}$ (h) $\frac{7}{9}$ (i) $\frac{9}{10}$
   8

3.

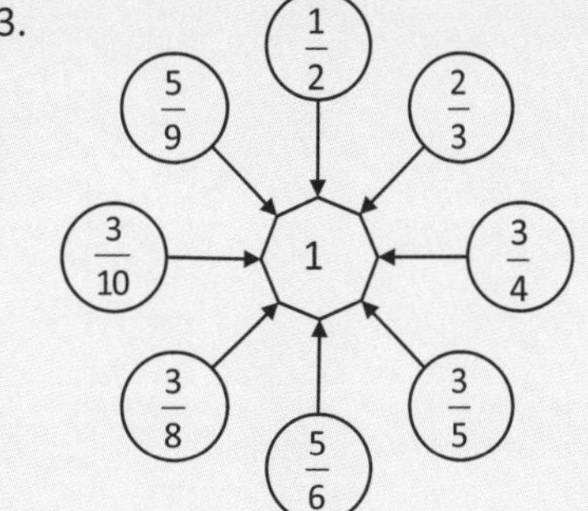

4. (a) $\frac{3}{5}$ (b) 1
   (c) $\frac{5}{8}$ (d) $\frac{7}{9}$
   (e) $\frac{6}{7}$ (f) 1
   (g) $\frac{3}{5}$ (h) $\frac{3}{4}$

Workbook

Exercise 10, pp. 112-114

1. (a) $\frac{3}{5}$ (b) $\frac{1}{6}$
   (c) $\frac{3}{8}$ (d) $\frac{3}{10}$
   (e) $\frac{1}{4}$ (f) $\frac{3}{5}$

2. (a) $\frac{1}{3}$ (b) $\frac{2}{5}$
   (c) $\frac{2}{3}$ (d) $\frac{5}{8}$
   (e) $\frac{1}{4}$ (f) $\frac{3}{4}$
   (g) $\frac{3}{5}$ (h) $\frac{3}{10}$
   (i) $\frac{1}{2}$ (j) $\frac{1}{12}$
   14

3.

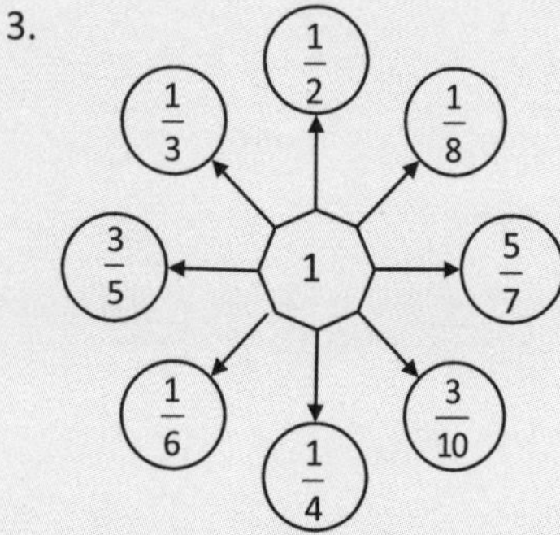

4. (a) $\frac{1}{2}$ (b) $\frac{2}{7}$
   (c) 0 (d) 0
   (e) $\frac{1}{2}$ (f) $\frac{1}{3}$
   (g) $\frac{3}{5}$ (h) $\frac{1}{3}$

Exercise 11, pp. 115-116

1.

12	12	12	12	12	12	12	12
0	1	2	4	6	8	10	12
$\frac{0}{12}$	$\frac{1}{12}$	$\frac{2}{12}$	$\frac{4}{12}$	$\frac{6}{12}$	$\frac{8}{12}$	$\frac{10}{12}$	$\frac{12}{12}$
0	$\frac{1}{12}$	$\frac{1}{6}$	$\frac{1}{3}$	$\frac{1}{2}$	$\frac{2}{3}$	$\frac{5}{6}$	1

2. (a) $\frac{2}{7}$ (b) $\frac{2}{5}$
   (c) $\frac{3}{4}$ (d) $\frac{1}{2}$

3. (a) $\frac{1}{3}$ of 15 = 15 ÷ 3 = **5**
   $\frac{2}{3}$ of 15 = 5 x 2 = **10**

   (b) $\frac{1}{4}$ of 16 = 16 ÷ 4 = **4**
   $\frac{3}{4}$ of 16 = 4 x 3 = **12**

   (c) $\frac{1}{5}$ of 20 = 20 ÷ 5 = **4**
   $\frac{2}{5}$ of 20 = 4 x 2 = **8**

   (d) $\frac{1}{6}$ of 12 = 12 ÷ 6 = **2**
   $\frac{5}{6}$ of 12 = 2 x 5 = **10**

   (e) $\frac{1}{8}$ of 16 = 16 ÷ 8 = **2**
   $\frac{5}{8}$ of 16 = 2 x 5 = **10**

   (f) $\frac{1}{10}$ of 20 = 20 ÷ 10 = **2**
   $\frac{3}{10}$ of 20 = 2 x 3 = **6**

Workbook

Exercise 12, p. 117

1. (b) \$0.75 $\frac{3}{4}$ of a dollar

   (c) \$0.50 $\frac{1}{2}$ of a dollar

   (d) \$0.75 $\frac{3}{4}$ of a dollar

   (e) \$0.40 $\frac{2}{5}$ of a dollar

   (f) \$1.00 $\frac{1}{1}$ of a dollar

Review 10, pp. 118-122

1. (a) 185 (b) 53 (c) 6700
   (d) 2972 (e) 3654 (f) 2304

2. Which part is colored can vary.

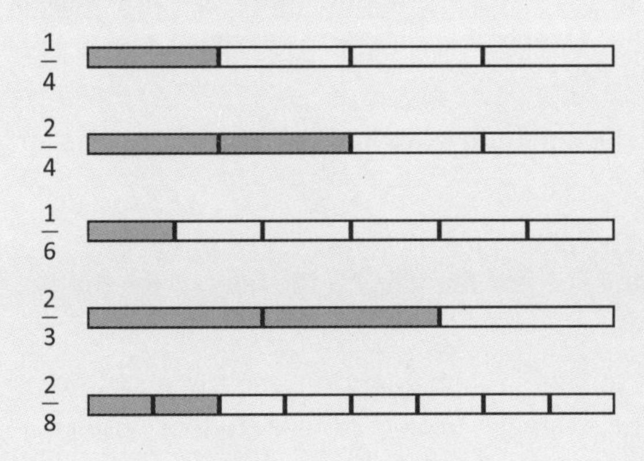

3. (a) $\frac{1}{4}$ (b) $\frac{3}{8}$ (c) $\frac{2}{3}$
   (d) $\frac{4}{6}$ (e) $\frac{3}{4}$ (f) $\frac{3}{8}$

4. \$8.55 x 3 = \$25.65
   Sean has **\$25.65**.

5.

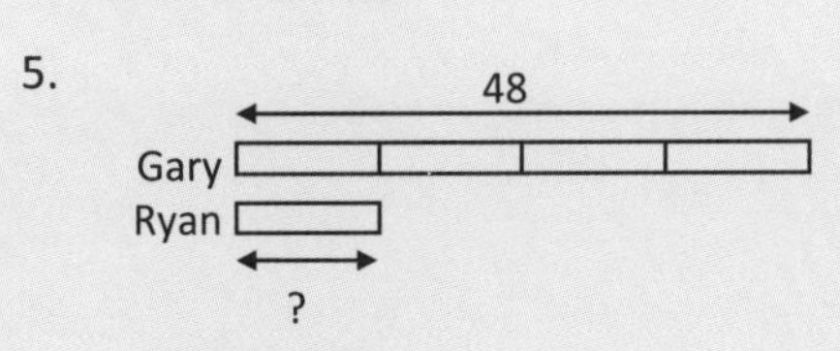

4 units = 48
1 unit = 48 ÷ 4 = 12
Ryan collected **12** cards.

6. 1 lb – 9 oz = 7 oz
   She had **7 oz** of flour left.

7.

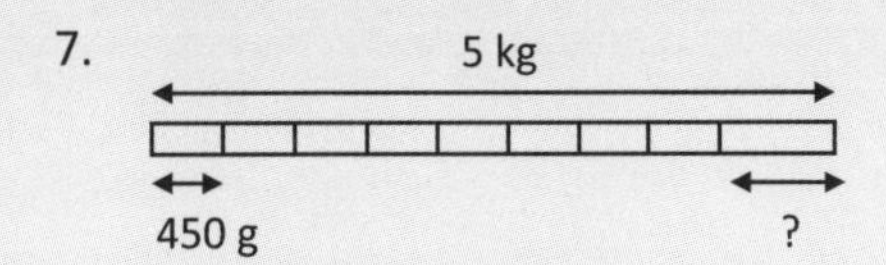

Weight of 8 bags of onions:
450 g x 8 = 3600 g = 3 kg 600 g
Weight of potatoes:
5 kg – 3 kg 600 g = 1 kg 400 g
The potatoes weigh **1 kg 400 g**.

8. (a) 1 roll: \$5
   4 rolls: \$5 x 4 = \$20
   The 4 rolls cost **\$20**.
   (b) 4 rolls make 10 bows.
   1 bow: \$20 ÷ 10 = \$2
   It costs **\$2** to make 1 bow.

9. Number of pies packed:
   157 – 37 = 120
   Group by 8's:
   120 ÷ 8 = 15
   There were **15** boxes of pies.

10. Put 12 ribbons into groups of 3.

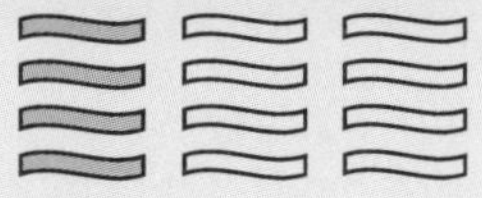

12 ÷ 3 = 4
Each group has 4 ribbons.
One group has red ribbons, the other 2 have yellow ribbons. 2 x 4 = 8

(a) $\mathbf{\frac{2}{3}}$ of her ribbons are yellow.

(b) **8** of her ribbons are yellow.

11. (a) 6 + 9 + 16 + 12 = **43**
    (b) **C**

12. (a) 21
    (b) 7
    (c) 14
    (d) 3

Unit 11 – Time

Chapter 1 – Hours and Minutes

Objectives

- ♦ Review telling time on an analog clock to the 5-minute mark.
- ♦ Review a.m. and p.m.
- ♦ Tell time on an analog clock to the 1-minute mark.
- ♦ Find elapsed time using clock faces.
- ♦ Convert between hours and minutes.
- ♦ Find the duration of a time interval.
- ♦ Find the ending time given the start time and the duration.
- ♦ Find the starting time given the duration and ending time.
- ♦ Solve word problems involving time.

Notes

In *Primary Mathematics* 2B, students learned to tell time to the 5-minute mark, to use a.m. and p.m., and to find the duration of time intervals or the starting or ending times using an analog clock face.

In this chapter, students will learn to tell time to the 1-minute mark and solve problems involving time without the use of a clock face.

Many of the problems involving time can be solved by counting up the hours and then the minutes. You can help your student by drawing a time line and marking off the hours from the start time. Do not require your student to draw time lines.

Students will learn to convert hours and minutes to minutes and minutes to hours and minutes, and then to add and subtract in compound units. The strategies used are similar to those used in adding and subtracting in compound units in the measurement chapters, except that now the conversion factor is 60.

The origin of the minute and second goes back to the Babylonians. The Babylonians did their astronomical calculations in the sexagesimal, or base-60 system. In the base-60 system, one place value can hold 59 units. When another unit is added, the 60 units are regrouped to the next place value. So each place value is 60 times the next lower place value, or each place value is one sixtieth of the next higher place value. The Babylonians divided the hour up into base-60 fractions. The first fractional sexagesimal place (one sixtieth of one whole) we now call a minute; the second place, one sixtieth of one sixtieth, we now call a second.

Material

- ♦ Two analog clocks with geared hands
- ♦ Large demonstration clock, if available
- ♦ Stopwatch
- ♦ Mental Math 12 (appendix)

(1) Tell time to the 1 minute interval

Activity

Provide your student with a geared clock and use another one yourself or a larger demonstration clock. Review telling time to the 5-minute interval if needed. Suggestions for review follow; if your student can already tell time on a clock face to the 5-minute interval you can skip them.

⇒ Make sure your student has a feel for how long an hour and a minute is. You can ask her for some activities that take about an hour. Have her estimate how long some activities, like getting dressed or brushing teeth, take in minutes. Use a stopwatch and ask her to do some activity like hopping for a minute, or do task 1 on p. 113 of the textbook. Then ask her to sit still and not talk or move for a minute. Discuss the fact that our perception of passing time is subjective; time seems to go faster when we are not paying attention to it, or doing something fun, than when we are bored or waiting for the time to pass.

⇒ Let your student experiment with the clock to observe the movement of the two hands and then explain to you how they move relative to each other. As an hour passes the longer minute hand moves all the way around the clock while the shorter hour hand moves from one number to the next. The hour hand measures the hours that have passed, and the minute hand measures the minutes that have passed. Ask her what the numbers on the clock mean. They mark the hour.

⇒ Ask your student how much time passes as the minute hand moves from one small mark to the next, and how much time passes as the minute hand moves from one number to the next. When the minute hand moves from one mark to the next, a minute has passed. There are 5 minutes from one number to the next. Have him count by 5's as you move the minute hand from 12 all the way around. Ask him for the number of minutes in an hour. There are 60 minutes in an hour.

⇒ Set some times to the 5-minute intervals on the clock and ask your student to tell you the time. Include quarter and half-hours, and telling time as both after the hour and before the hour for times after the half hour. Get him to write the times in digital notation, i.e. hours:minutes.

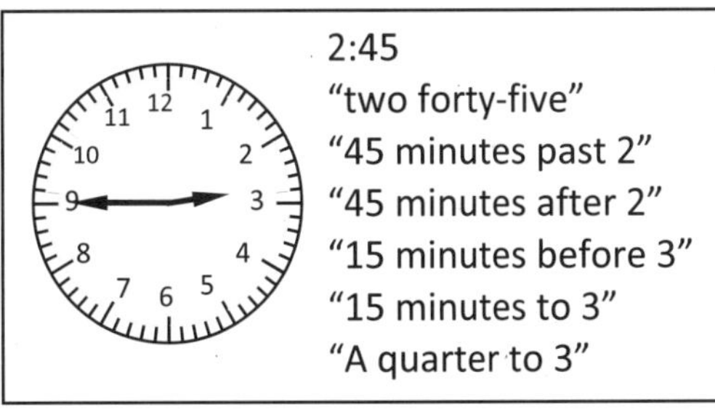

Review the 24 hour day and a.m. and p.m., if needed:

⇒ Show the time 12:00. Tell your student that this is 12 midnight, in the middle of the night. 12 midnight is considered to be the start of one full "day" which includes both night time and day time. Have your student show the times for various activities done in the morning, and then on through the day, by moving just the minute hand, going through 12 noon, and then moving the hand past her bed-time to 12 midnight again.

⇒ Remind your student that we use a.m. to stand for the times between 12:00 midnight and 12:00 noon and we use p.m. to stand for the times between 12:00 noon and 12:00 midnight.

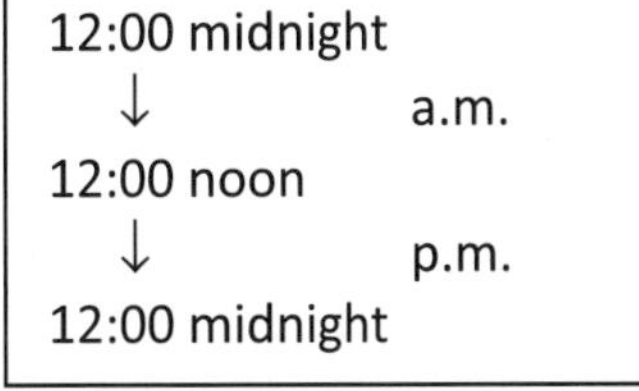

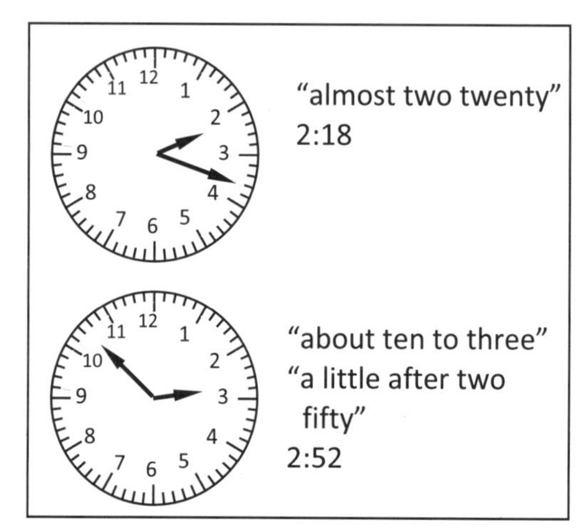

Set the time on a geared clock to 2:18. Have your student first estimate the time to the nearest 5 minutes. It is almost 2:20. Then have her find the exact time. She can count by 5's to 15 and then on by 1's, or recognize that the time is a little past a quarter after the hour and just count on by 1's from 15.

Set the time to 2:52. Get your student to first estimate the time to the nearest 5 minutes and then find the exact time. He can count up from 3-quarters of an hour, or 2:45, first to 2:50 and then to 2:52.

Repeat with other examples as necessary.

Point out that we do not always have to know the time to the exact minute, and so simply estimating the time to the nearest 5 minutes is often sufficient.

Discussion

Concept pages 112-113

Have your student set the start time 8:20 on his clock and then move the minute hand to the end time, counting the minutes by 5's to see that 15 minutes have passed.

Practice

Task 2, p. 113

2. (a) 2:05 (b) 4:15
(c) 12:20 (d) 7:30
(e) 3:43 (f) 7:52

Workbook

Exercise 1, pp. 123-124 (Answers p. 125)

For problem 2, the exact words used are not important, but you can tell your student to use "minutes past" for times up to the half hour, and "minutes to" for all times after the half hour, for example, 18 minutes **to** 5.

Reinforcement

Dictate a time as minutes past the hour or write the time down and let your student set the corresponding time on the clock.

(2) Find elapsed time using clock faces

Discussion

Tasks 3-4, p. 114

> 3. 9:56
>
> 4. 60 minutes
> 4:00 to 5:00
> 6:45 to 7:45

3: There are several ways your student can find the end time. 30 + 26 is 56, so it is 9:56. Or, he can see that 26 minutes is 4 minutes short of another half hour, and 60 – 4 = 56.

4: Have your student tell you both the starting time and ending times in these two tasks. Point out that the minutes stay the same and only the hour changes. The minute hand moves around once to the same place it started.

Activity

Write down a starting time of 11:15 a.m. and an ending time of 3:15 p.m. for some activity such as a car drive. Ask your student how much time has passed. Since the minutes stay the same, we can simply count up the hours, but we have to switch after 12 back to 1. You can show this switch on the clock. When we go past 12, a.m. switches to p.m.

Draw a number line like the one at the right. Tell your student that we can think of time passing as a time line. The start of the time line is the starting time. On this one, each mark is the next hour.

start 11:15 a.m. end 3:15 p.m.
(11:15) 12, 1, 2, 3:15
4 hours

11:15 12:15 1:15 2:15 3:15
4 h

Tell your student that instead of arriving at 3:15 p.m., we get to our destination at 3:52 p.m. Ask him how long the drive took. We can still count up by hours to 3:15, and then we find the number of minutes from 3:15 to 3:52. Using the clock face, we can see that it is 30 minutes to go half-way around to 3:45, and another 7 minutes to 3:52. So the drive took 4 hours and 37 minutes. Show this on the time line, adding another hour and then an intermediate mark for 3:52.

start 11:15 a.m. end 3:52 p.m.
(11:15) 12, 1, 2, 3:15, 3:45, 3:52
4 hours 37 minutes

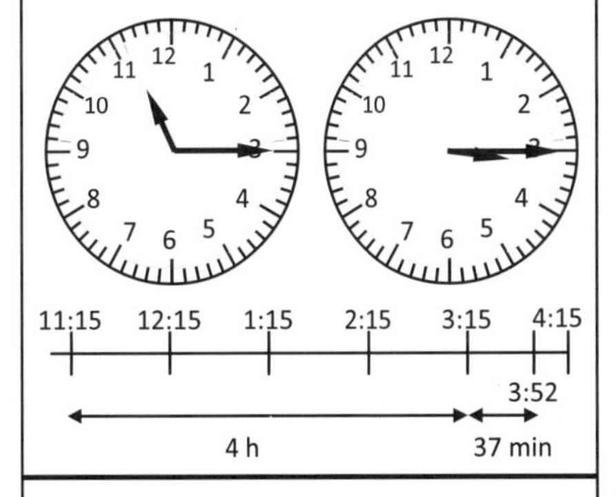

Ask your student to find how long it is from 10:50 p.m. to 2:14 a.m. She can use the clock face. This time, p.m. goes to a.m. as we count up 3 hours to 1:50. There is another 10 minutes to 2:00, and then 14 more minutes to 2:14. So 3 hours 24 minutes have passed.

start 10:50 p.m. end 2:14 a.m.
(10:50) 11, 12, 1:50, 2:00, 2:14
3 hours 24 minutes

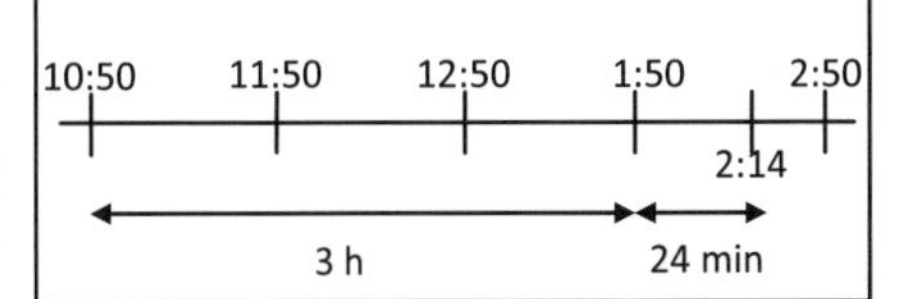

Practice

Task 5, p. 115

> 5. (a) 27 min (b) 5 hours
> (b) 2 h 15 min (d) 2 h 30 min

Workbook

Exercise 2, pp. 125-126 (Answers p. 125)

(3) Convert between hours and minutes

Activity

Ask your student for the number of minutes in 1 hour, 2 hours, and so on to 10 hours. Write the answers. To find the number of minutes, we can multiply the number of hours by 60 (6 tens, multiply by 6 and then 10).

Ask your student how many hours are in 180 minutes, 480 minutes, and 540 minutes. To find the number of hours, we can divide by 60 (divide by 10 and then 6).

1 hour	→	60 minutes
2 hours	→	120 minutes
3 hours	→	180 minutes
4 hours	→	240 minutes
5 hours	→	300 minutes
6 hours	→	360 minutes
7 hours	→	420 minutes
8 hours	→	480 minutes
9 hours	→	540 minutes
10 hours	→	600 minutes

Discussion

Tasks 6-7, p. 116

6: To find who took the longest or shortest time, we first compare the hours, and if they are the same, we compare the minutes.

7: To convert hours and minutes to minutes, we find the number of minutes in the hours, and then add on the rest of the minutes.

6. (a) **Jane** (2 h > 1 h)
   (b) **Amy** (15 min < 20 min)

7. 60 + 35 = **95** min

Practice

Task 8, p. 116

8. (a) 120 min (b) 130 min (c) 165 min
   (d) 180 min (e) 185 min (f) 195 min

Activity

Ask your student how many hours are in 405 minutes. We want to find the number of hours with a remainder of minutes. We can round down to the ten, 400, and then divide 40 tens by 6 tens by thinking of the answer to 40 ÷ 6. 36 is the closest number below 40 that can be divided by 6, and 36 is 6 x 6. So there are 6 hours and some minutes in 405 minutes. To then find the number of minutes left, we subtract the number of minutes in 6 hours, 360, from 405. We can do this mentally; 360 is 40 more to 400, add 5, there 45 minutes left over. So 405 is 6 hours 45 minutes.

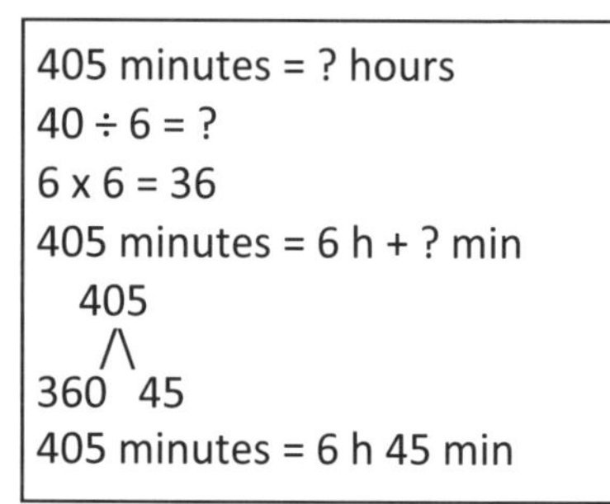

Discussion

Task 9, p. 116

9. 180 min = 3 h
   200 min = 3 h 20 min

Practice

Task 10, p. 116

10. (a) 1 h 10 min (b) 1 h 25 min (c) 1 h 40 min
    (d) 2 h 5 min (e) 2 h 40 min (f) 3 h 30 min

Workbook

Exercise 3, pp. 127-128 (Answers p. 125)

(4) Solve word problems involving time

Discussion

11. 1 h 5 min
12. 9:00 a.m.
13. 8:30 p.m.

Tasks 11-1, p. 117

11: 8 to 9 is one hour, then another 5 minutes to 9:05

12: 7:15 to 8:15 is 1 h. Another 45 minutes would move the minute hand to 9:00. Your student should remember that 15 minutes is a quarter of an hour, and 45 minutes is three quarters of an hour, so adding another 45 min to 15 min would fill out the hour.

13: 1 hour before 9:40 is 8:40, and another 10 minutes sooner is 8:30.

Activity

Ask your student to solve the following problems. You can draw timelines to help your student with counting hours and minutes.

⇒ Joe arrived at the zoo at 9:40 a.m. and stayed until 3:20 p.m. How much time did he spend at the zoo?

We can count up the hours from 9:40 to 2:40. We need to remember to switch from 12 to 1 when counting from 12:40 to 1:40. Then there are 20 minutes to 3:00 and another 20 minutes to 3:20. He spent 5 h 40 min at the zoo.

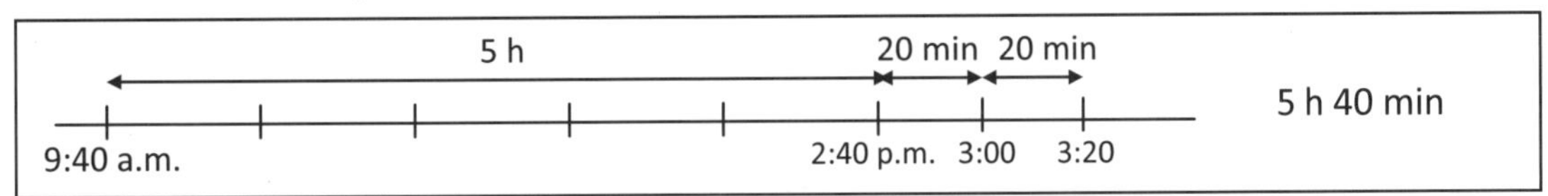

⇒ Paul went to a program on reptiles at the zoo. He arrived at 11:50 a.m. The program ended 3 hours and 25 minutes later. When did the program end?

Count up 3 hours to 2:50, add 10 of the minutes to 3:00, and then the 15 minutes left. The program ended at 3:15 p.m.

⇒ Mr. Ahroni works at the zoo for 8 h 30 minutes every day. He finishes work at 3:15 p.m. At what time does he start work?

Go back 8 hours: 3 hours from 3 back to 12:15, another 5 hours back to 7:15 a.m., then back another half hour to 6:45 a.m. He starts work at 6:45 a.m.

⇒ It takes 1 h 20 minutes to feed the seals and 3 h 15 min to feed all the monkeys. How much longer does it take to feed the monkeys than the seals?

We can solve this by counting on as well, as if the time line started at 12 and we need to find the duration between 1:20 and 3:15. We can either count on 1 h to 2:20 min, 40 min to 3:00, 15 more min to 3:15, for a total of 1 h 55 min. Or, count on 2 h to 3:20 and back 5 minutes. It takes 1 h 55 min longer to feed the monkeys.

Workbook

Exercise 4, pp. 129-130 (Answers p. 125)

Reinforcement

Extra Practice, Unit 11, Exercise 1A, pp. 197-200

(5) Add and subtract time in hours and minutes

Discussion

Ask your student to subtract some minutes from an hour or several hours. One method is to imagine a clock face and the amount of time from the minutes given to the next hour. A second method is to count up to 60 by tens and then ones. A third method is to subtract tens and then ones from 60.

⇒ 1 h – 10 min

⇒ 1 h – 40 min

⇒ 2 h – 40 min

⇒ 3 h – 25 min

```
1 h – 10 min = 50 min

1 h – 40 min = 20 min

   2 h – 40 min = 1 h 20 min
  /\
1 h  60 min

3 h – 25 min = 2 h 35 min
```

Give your student the following problems. He can either add the minutes and then convert the total minutes to hours, or he can add by making an hour with one set of minutes. This is similar to what he has already been doing to find elapsed time.

⇒ 30 min + 50 min

⇒ 25 min + 45 min

⇒ 1 h 25 min + 45 min

⇒ 6 h 45 min + 20 min

```
30 min + 50 min
          /\
        30  20
or
  30 min + 50 min = 1 h 20 min
  /\
20  10

25 min + 45 min = 1 h 10 min
          /\
        35  10

1 h 25 min + 45 min = 2 h 10 min

6 h 45 min + 20 min = 7 h 5 min
              /\
            15  5
```

Give your student the following problems. To subtract the minutes, when there are not enough minutes to subtract from, we can either convert 1 h and the minutes first and then subtract, or we can subtract the minutes from an hour and add the difference.

⇒ 1 h 15 min – 45 min

⇒ 1 h 5 min – 30 min

⇒ 3 h 5 min – 30 min

⇒ 3 h 20 min – 55 min

```
  1 h 15 min – 45 min
 /
1 h – 45 min = 15 min
                |
               15 min + 15 min= 30 min
  Or
  1 h 15 min
     |
  75 min – 45 min = 30 min

1 h 5 min – 30 min = 30 min + 5 min
                   = 35 min

  3 h 5 min – 30 min = 2 h 35 min
  /\
2 h  1 h

3 h 20 min – 55 min = 2 h 5 min + 20 min
                    = 2 h 25 min
```

Discussion

Tasks 14-17, pp. 118-119

These tasks show an alternate way of finding a time duration for times that go from a.m. to p.m. or p.m. to a.m. In the previous lesson your student counted up the hours and minutes, going from 12 to 1 when passing noon or midnight, to find the time duration. In this lesson she will find the time from the starting time to 12:00, then from 12:00 to the ending time, and then add the two time durations together.

15: To add time in compound units, we first add the hours and then the minutes.

17: To subtract time in compound units we first subtract the hours and then the minutes.

14. (a) 2 h
    (b) 3 h 30 min
    (c) 1 h 15 min

15. 11 h 15 min
    11 h 15 min

16. (a) 4 h
    (b) 6 h 40 min
    (c) 2 h 50 min

17. 1 h 50 min
    1:50 a.m.

Practice

Task 18, p. 119

18. (a) 5 h 40 min (b) 3 h 5 min
    (c) 1 h 15 min (d) 2 h 35 min
    (e) 3 h 40 min (f) 5 h 5 min
    (g) 2 h 15 min (h) 5 h

Workbook

Exercise 5, pp 131-132 (Answers p. 126)

Reinforcement

Mental Math 12

Extra Practice, Unit 11, Exercise 1B, pp. 201-204

(6) Practice

Practice

Practice A, p. 120

Test

Tests, Unit 11, 1A and 1B, pp. 201-208

Enrichment

Tell your student that in some countries and in the military they use a 24 hour clock in which the day from midnight to midnight is divided up into 24 hours rather than two 12 hour segments. 1:00 p.m. is written 13:00 and read as "thirteen hundred." The last minute of the day is 23:59; at the end of the next minute it is again 00:00. 24:00 is mainly used to refer to the exact end of a day in a time interval, such as 21:00-24:00 for 9:00 p.m. to midnight. Usually a 0 is used for the first digit for times less than 10:00, such as 07:15. This can be read as "oh seven fifteen." The minutes for times on the hour are often read as "hundred" even though there are not 100 minutes in an hour. For example, 09:00 is "oh nine hundred."

1. (a) 3 h 45 min (b) 1 h 40 min
   (c) 3 h (d) 1 h 15 min
   (e) 3 h 20 min (f) 40 min
2. 2:25
3. (a) 7 h 15 min
   (b) 1 h 15 min
   (c) 1 h 55 min
   (d) 45 min
4. (a) 2 h 35 min + 1 h 55 min = 4 h 30 min
   He took **4 h 30 min** to repair both vehicles.
   (b) 2 h 35 min – 1 h 55 min = 40 min
   He took **40 min** longer to repair the van.
5. He finished at **11:50 a.m.**
6. They returned at **12:40 p.m.**
7. They must report for work at **8:50 a.m.**

Have your student convert some times to the 24-hour clock. For p.m. times, it would be 12 plus the hours:

⇒ 12:30 a.m.
⇒ 4:15 a.m.
⇒ 1:30 p.m.
⇒ 3:00 p.m.
⇒ 6:00 p.m.
⇒ 9:40 p.m.

12:30 a.m. → 00:30
4:15 a.m. → 04:15
1:30 p.m. → 13:30
3:00 p.m. → 15:00
6:00 p.m. → 18:00
9:40 p.m. → 21:40

Have your student convert some times from the 24-hour clock to the 12-hour clock, using a.m. and p.m. For times after 12:00 noon, subtract 12 from the hours.

⇒ 00:45
⇒ 07:15
⇒ 12:45
⇒ 14:00
⇒ 17:40
⇒ 20:30

00:45 → 12:45 a.m.
07:15 → 7:15 a.m.
12:45 → 12:45 p.m.
14:00 → 2:00 p.m.
17:40 → 5:40 p.m.
20:30 → 8:30 p.m.

Chapter 2 – Other Units of Time

Objectives

- Understand the magnitude of a second relative to a minute.
- Convert between seconds and minutes and seconds
- Understand the relationships between days, weeks, months, and years.

Vocabulary

- Seconds

Notes

In this chapter, students will measure time in seconds, and convert minutes and seconds to seconds, and seconds greater than 60 to minutes and seconds. Since there are 60 seconds in a minute, as there are 60 minutes in an hour, the process is similar to that used to convert between hours and minutes.

Races are often measured in seconds or minutes and seconds. Make sure your student understands that a person who is faster takes *less* time than one that is slower, and the *fastest* person has the *least* time.

In *Primary Mathematics* 2B, students learned to convert between days, weeks, months, and years. This is reviewed in this chapter.

⇒ 1 day = 24 hours

⇒ 1 week = 7 days

⇒ 1 month = about 4 weeks (28-31 days)

⇒ 1 year = 12 months

⇒ 1 year = about 52 weeks

⇒ 1 year = 365 days (366 days on leap year)

Material

- Stopwatch
- Calendar
- Mental Math 13-14 (appendix)

(1) Measure time in seconds

Discussion

Ask your student how we measure times that are less than a minute. We use seconds.

If you have a wall clock with a second hand, have your student observe the movement of the second hand. If you have a digital clock or timer with seconds, such as on the microwave, or a stopwatch, your student can observe the change in the numbers as the seconds change. From this, she should be able to determine that there are 60 seconds in a minute. Tell her that time can be written with seconds by using another colon. For example, 02:30:45 is 2 hours 30 minutes 45 seconds. Seconds will be abbreviated as the letter "s".

Discussion

Concept p. 121

Tasks 1-2, pp. 121-122

You can do these suggested tasks or measure other activities in seconds to get your student familiar with seconds. Make a table of the activities and the times needed to do those activities. Then ask him which activity was done the fastest. It would be the activity with the least time. Have him order the activities from fastest to slowest.

Activity

Ask your student whether we would measure various activities, such as the following, in hours, minutes, or seconds. Use activities she is familiar with.

- ⇒ Take a bath (minutes)
- ⇒ Wash and dry your hands (seconds or minutes)
- ⇒ Brush your hair (seconds or minutes)
- ⇒ Say your name (seconds)
- ⇒ Sleep at night (hours)
- ⇒ Work on homework (minutes or hours)
- ⇒ Make and bake bread (hours)
- ⇒ Play a game of soccer (hours)

Discussion

Task 3, p. 123

Your student can follow the same process in converting from minutes and seconds to seconds, or from seconds to minutes and seconds, as was used in the previous chapter for converting between minutes and hours.

3. (a) 180 s + 40 s = **220 s**
   (b) 2 min 30 s

Workbook

Exercise 6, pp. 133-136 (Answers p. 126)

(2) Convert between years and months

Activity

If necessary, review the organization of years into months. Use a calendar. Have your student count the number of months in a year, name them, and tell what season or seasons they fall in. Discuss particular events or holidays in specific months that occur year after year.

Have your student use the calendar to examine the number of days in each month. You may want to tell him that a day is how long it takes for the earth to make one revolution around its axis, and that a year is the time the earth goes around the sun and is not exactly an even number of days. There are 365 days in a year, but it takes the earth about another fourth of a day to go all the way around the sun. So to keep the seasons at the same time of the year over a long period of time, we add one day to the calendar every four years in February. Occasionally, additional time needs to be added; for example, in 2008 a second was added to the year. A year with 366 days is called a leap year. So usually February has 28 days, but every 4 years it has 29 days. The rest of the months have 30 or 31 days.

Ask your student for the number of months in 1, 2, 3, years and so on up to 10 years. To find the number of months given the number of years, we multiply the years by 12. Your student should be comfortable with the multiplication facts for 12 by now.

1 year → 12 months
2 years → 12 months x 2 = 24 months
3 years → 12 months x 3 = 36 months
4 years → 12 months x 4 = 48 months
5 years → 12 months x 5 = 60 months
6 years → 12 months x 6 = 72 months
7 years → 12 months x 7 = 84 months
8 years → 12 months x 8 = 96 months
9 years → 12 months x 9 = 108 months
10 years → 12 months x 10 = 120 months

Ask your student how we would find the number of years, given the number of months greater than or equal to 12. We divide by 12. The quotient is the number of years, and the remainder is the number of months left over. Ask your student for the number of years and months in 80 months.

80 months = ? years
6 x 12 = 72   80
/\
72 8
80 months = 6 years 8 months

Practice

Task 4, p. 123

4. (a) 12 months
(b) 24 months
(c) 28 months (24 + 4)
(d) 3 yrs 4 months   40
/\
36 4

Workbook

Exercise 7, pp. 137-138 (Answers p. 126)

Enrichment

Have your student hold out his two fists in front of himself as shown here. We can assign the months in order to each knuckle and valley between the knuckles from left to right. If a month falls on a knuckle, it has 31 days, if it falls on a valley, it has 30 days, except for February.

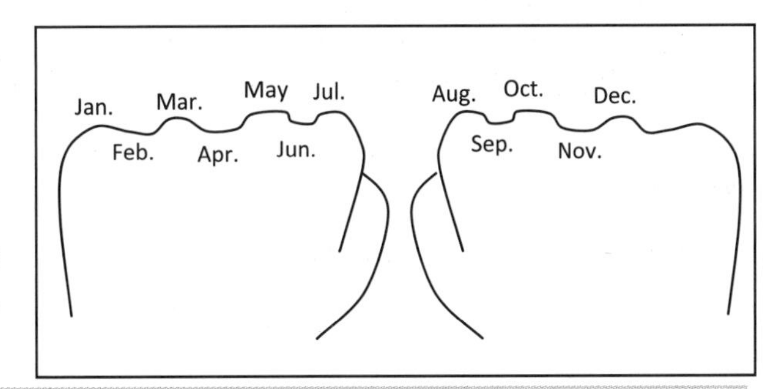

(3) Convert between weeks and days

Activity

Refer to a calendar page. Discuss the names of the days of the week and ask your student how many days there are in a week. You may want to discuss how the days progress from one month to the next; the number of the day starts over at the beginning of the next month and the first day of the next month is the day of the week after the last day of the previous month. Make sure your student understands that a week is 7 days and it does not matter what day the week starts on, even though we usually think of Sunday as the first day in the week since it is the first day on the calendar (though some calendars have Monday first). 1 week from Wednesday is the next Wednesday, which is 7 days later.

Ask your student how many days there are in 2 to 10 weeks. Point out that since 1 week is 7 days, we multiply the number of weeks by 7 to find the number of days.

1 week → 7 days
2 weeks → 7 days x 2 = 14 days
3 weeks → 7 days x 3 = 21 days
4 weeks → 7 days x 4 = 28 days
5 weeks → 7 days x 5 = 35 days
6 weeks → 7 days x 6 = 42 days
7 weeks → 7 days x 7 = 49 days
8 weeks → 7 days x 8 = 56 days
9 weeks → 7 days x 9 = 63 days
10 weeks → 7 days x 10 = 70 days

Ask your student how we can find the number of weeks if we are given the number of days, such as 50 days. We divide by 7. The quotient is the number of weeks, and the remainder the left over days. Ask your student to find the number of weeks and days in 50 days.

50 days = ? weeks
7 x 7 = 49 50
/\
49 1
50 days = 7 weeks 1 day

Ask your student how many weeks there are in a non-leap year. There are 52 weeks and 1 day.

365 days = ? weeks
365 ÷ 7 = 52 R 1
365 days = 52 weeks 1 day

Ask your student how many weeks there are in a month. Since a month is a bit more than 4 weeks long, we can only estimate the number weeks for a given number of months. 2 months is about 8 weeks and a half, 3 months is about 13 weeks.

1 month = ? weeks
1 month = a bit more than 4 weeks.

Practice

Task 5, p. 123

5. (a) 7 days
(b) 21 days
(c) 25 days (21 + 4)
(d) 4 weeks 2 days 30
/\
28 2

Workbook

Exercise 8, pp. 139-140 (Answers p. 127)

Reinforcement

Extra Practice, Unit 11, Exercise 2, pp. 205-206

(4) Practice

Practice

Practice A, p. 124

1. (a) 132 min (b) 1 h 48 min
   (c) 123 s (d) 1 min 34 s
   (e) 21 months
   (f) 2 years 6 months
   (g) 19 days
   (h) 5 weeks 5 days
2. 3 h 15 min – 1 h 35 min = 1 h 40 min
   It takes **1 h 40 min** longer to fly to Miami.
3. The bookshop is open for **7 h 30 min**.
4. She returned home at **2:20 p.m.**
5. Lily took **56 min** to complete the puzzle.
6. The journey took **45 min**.
7. He left Los Angeles at **5:30 a.m.**

Reinforcement

Mental Math 13

Have your student add and subtract in compound units for years and months and for weeks and days.

- ⇒ 1 week – 4 days (3 days)
- ⇒ 3 weeks – 4 days (2 weeks 3 days)
- ⇒ 3 weeks 2 days – 4 days (2 weeks 5 days)
- ⇒ 3 weeks 2 days – 1 week 4 days (1 week 5 days)
- ⇒ 1 year – 7 months (5 months)
- ⇒ 6 years – 7 months (5 years 5 months)
- ⇒ 6 years 2 months – 7 months (5 years 7 months)
- ⇒ 6 years 2 months – 5 years 7 months (7 months)

Test

Tests, Unit 11, 2A and 2B, pp. 209-214

Enrichment

Ask your student how many seconds there are in an hour.

Mental Math 14

1 hour = ? seconds
1 hour = 60 minutes
60 minutes = 60 x 60 seconds
= 60 x 6 x 10
= 3600 seconds
1 hour = 3600 seconds

Review 11

Review

Review 11, pp. 125-126

Workbook

Review 11 pp. 141-145 (Answers p. 127)

Test

Tests, Units 1-11, Cumulative A and B, pp. 215-226

1. (a) 8:55 p.m.
   (b) 1:30 a.m.
2. (a) $\frac{\mathbf{4}}{6}$ (b) $\frac{6}{\mathbf{8}}$ (c) $\frac{\mathbf{8}}{10}$
3. (a) 4 out of 8 = $\frac{1}{2}$
   (b) 2 out of 10 = $\frac{1}{5}$
4. 2 years is 24 months, which is 5 more than 19.
   Mr. Lee stayed 5 + 4 = **9 months** longer.
5. 10 minutes: 140 jars
   1 minute: 140 ÷ 10 = 14 jars
   It can fill **14** jars in 1 minute.
6. The lesson ended at **6:25 p.m.**
7. 60 ℓ
   Tank
   Bucket
   10 units = 60 ℓ
   1 unit = 60 ℓ ÷ 10 = 6 ℓ
   The capacity of the bucket is **6 ℓ.**
8. 1 m − $\frac{5}{8}$ m − $\frac{2}{8}$ m = $\frac{1}{8}$ m
   The third piece is $\frac{\mathbf{1}}{\mathbf{8}}$ **m** long.
9. \$6.50 + \$1.80 = \$8.30
   He paid **\$8.30**.
10. Number given to children:
    16 x 3 = 48
    Number in box: 48 + 20 = 68
    There were **68** mangoes in the box.
11. \$4.80 + \$2.50 = \$7.30
    It cost her **\$7.30**.
12. (a) \$40 − \$7.60 = \$32.40
    8 towels cost **\$32.40**.
    (b) \$32.40 ÷ 8 = \$4.05
    1 towel costs **\$4.05**.

Workbook

Exercise 1, pp. 123-124

1.

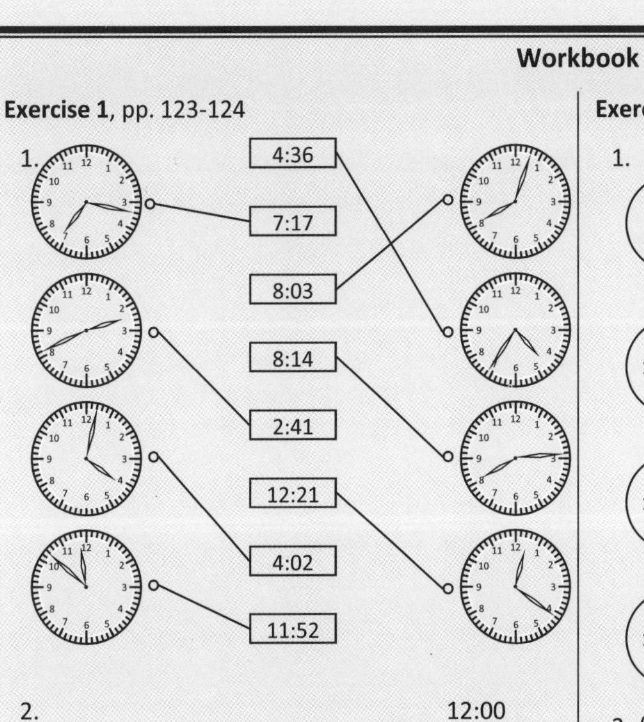

2.

		12:00 noon/midnight/12
4:42 18 minutes to 5	9:10 10 minutes past 9	2:45 15 minutes to 3
4:55 5 minutes to 5	11:05 5 minutes past 11	1:27 27 minutes past 1
7:25 25 minutes past 7	10:36 24 minutes to 11	8:53 7 minutes to 9

Exercise 2, pp. 125-126

1. (a) 3:15 p.m. — **25** minutes later — 3:40 p.m.
   (b) 5:35 a.m. — **30** minutes later — **6:05** a.m.
   (c) **6:10** p.m. — **2** hours later — **8:10** p.m.
   (d) **7:25** a.m. — **3** hours later — **10:25** a.m.

2. (a) 5:15 p.m. — **1** h **15** min later — 6:30 p.m.
   (b) **2:20** p.m. — **3** h **50** min later — **6:10** p.m.
   (c) **7:40** a.m. — **8** h **0** min later — **3:40** p.m.
   (d) **8:50** a.m. — **4** h **25** min later — **1:15** p.m.

Exercise 3, pp. 127-128

1.

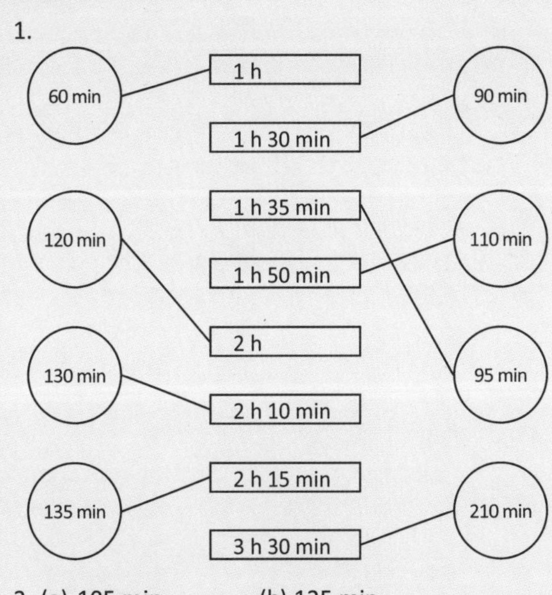

2. (a) 105 min (b) 125 min
   (c) 1 h 25 min (d) 2 h 30 min

3. (a) 65 min
   (b) 90 min
   (c) 145 min
   (d) 190 min

4. (a) 1 h 15 min
   (b) 1 h 40 min
   (c) 2 h 20 min
   (d) 3 h 45 min

Exercise 4, pp. 129-130

1. The show ended at **9:15 p.m.**
2. It took **1 h 20 min** for him to catch the first fish.
3. He reached the theater at **7:10 p.m.**
4. She waited for **1 h 20 min**.
5. He arrived at **8:20 a.m.**
6. Adam took **25 min** longer than James.

Workbook

Exercise 5, pp. 131-132

1. (a) 1 h 55 min (b) 2 h 25 min
   (c) 3 h 20 min (d) 3 h 10 min
2. (a) 4 h 10 min (b) 3 h 10 min
   (c) 4 h 10 min (d) 4 h 15 min
   (e) 5 h 10 min (f) 6 h 10 min
3. (a) 1 h 35 min (b) 2 h 25 min
   (c) 1 h 15 min (d) 2 h 55 min
4. (a) 1 h 20 min (b) 1 h 5 min
   (c) 1 h 15 min (d) 1 h 30 min
   (e) 1 h 45 min (f) 2 h 55 min

Exercise 6, pp. 133-136

1.

1 min	65 s
1 min 5 s	60 s
1 min 45 s	105 s
2 min	145 s
2 min 25 s	215 s
3 min	120 s
3 min 35 s	180 s

2. (a) 100 s (b) 130 s
   (c) 1 min 40 s (d) 2 min 30 s
3. (a) 85 s
   (b) 165 s
   (c) 170 s
   (d) 210 s
4. (a) 1 min 30 s
   (b) 1 min 55 s
   (c) 2 min 5 s
   (d) 3 min 20 s
5. (a) 90 s (b) 115 s
   (c) 125 s (d) 150 s
   (e) 185 s (f) 220 s
   (g) 1 min 20 s (h) 1 min 25 s
   (i) 1 min 35 s (j) 1 m 50 s
   (k) 2 m 20 s (l) 2 m 45 s
   A pentagon has 5 sides.
6. (a) Emily
   (b) Taylor
   (c) 4 s
7. (a) 20 s
   (b) 26 s
   (c) 45 s
   (d) 34 s

Exercise 7, pp. 137-138

1.

1 year 1 month	18 months
1 year 6 months	13 months
2 years	30 months
1 years 8 months	24 months
2 years 6 months	20 months
3 years	26 months
2 years 2 months	36 months

2. (a) 17 months (b) 2 years 4 months
3. (a) 15 months (b) 29 months
   (c) 35 months (d) 46 months
4. (a) 1 year 3 months
   (b) 2 years 1 month
   (c) 2 years 8 months
   (d) 3 years 4 months

Workbook

Exercise 8, pp. 139-140

1.

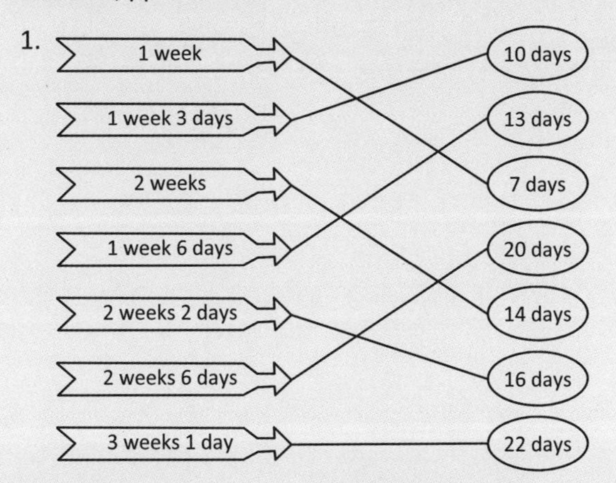

2. (a) 17 days (b) 5 weeks 5 days

3. (a) 12 days (b) 18 days
   (c) 24 days (d) 30 days

4. (a) 1 week 5 days
   (b) 3 weeks 4 days
   (c) 4 weeks 2 days
   (d) 4 weeks 4 days

Review 11, pp. 141-145

1. (a) $\frac{7}{10}$ (b) $\frac{5}{6}$
   (c) $\frac{3}{4}$ (d) $\frac{1}{2}$

2. (a) 10:25 a.m.
   (b) 30 minutes

3. (a) 4 (b) 30
   (c) 10 (d) 10

4. (a) m
   (b) ml
   (c) g
   (d) ℓ
   (e) km

5. (a) $10.45 – $8.60 = $1.85
   The storybook is $**1.85** cheaper.
   (b) $36.90 + $40.80 + $28.40 = $106.10
   He spent $**106.10**.
   (c) $40 – $8.95 – $8.60 – $10.45 = $12
   She received $**12** change.

6. 1 qt – 1 c = 3 c
   3 c of milk are left.

7. 4328 + 5860 = 10,188
   10,188 tickets were sold on both days.

8. 204 ÷ 3 = 68
   Each child got **68** stamps.

9. $20 – $1.25 – $12.50 = $6.25
   She received $**6.25** change.

10. $48.56
    watch
    calculator
    ?

    4 units = $48.56
    1 unit = $48.56 ÷ 4 = $12.14
    5 units = $12.14 x 5 = $60.70 ←
    Or: $48.56 + $12.14 = ~~$60.17~~
    The total cost is $~~**60.17**~~

11. (a) 10 x 28 = 280
    There were **280 children**.
    (b) 280 – 136 = 144
    There were **144 boys**.

12. (a) $45 + $60 + $50 + $65 + $40 + $35 = $**295**
    (b) April

Unit 12 – Geometry

Chapter 1 – Angles

Objectives

- Identify angles.
- Relate the size of angles to the degree of turning.
- Identify polygons.
- Relate the number of angles in a polygon to the number of sides.

Vocabulary

- Angle
- Polygon
- Triangle
- Quadrilateral
- Pentagon
- Hexagon
- Octagon

Notes

An angle is formed when two straight lines meet at a point. The point of intersection is the vertex of the angle, and the two lines are the sides of the angle.

The size of an angle is determined by how much either line is turned about the point where they meet. It does not depend on the length of the two sides.

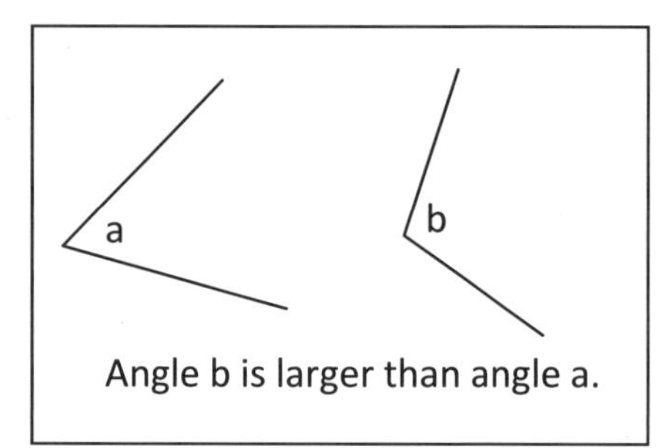

Angle b is larger than angle a.

A polygon is a closed figure formed by straight sides. Students will learn to identify polygons with 3, 4, 5, 6, and 8 sides by name. They will also learn that the number of angles is the same as the number of sides of a polygon.

Material

- Two cards, such as index cards fastened to each other along their width with tape.
- Folding meter stick or two strips of cardboard attached with a brad so they can rotate relative to each other, or something similar from a toy building set.

(1) Identify angles and polygons

Discussion

Concept p. 127

Have your student do this activity with two cards taped along their width. Point out the angle formed by the card by tracing it with your finger.

Activity

Use the cards or a folding meter stick or two strips of cardboard fastened with a brad at the bottom. Start with it closed up and then open to trace larger and larger angles. Tell your student that the larger the angle, the more one side has to be turned away from the other. The size of the angle depends on the degree of turning. Include an angle larger than a straight line. Point out that the size of the angle can go all the way around to a full circle (and beyond).

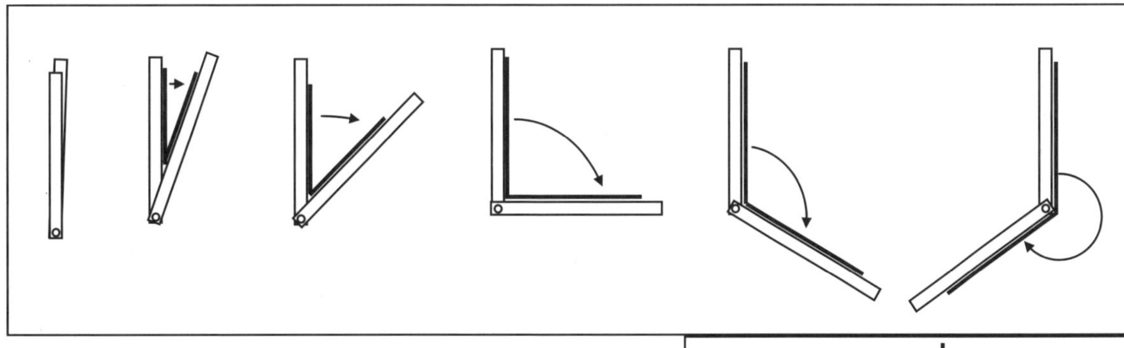

For one of the angles you have drawn, extend the arms. Ask your student if extending the arms makes the angle larger. It does not; the size of the angle does not depend on the length of the arms.

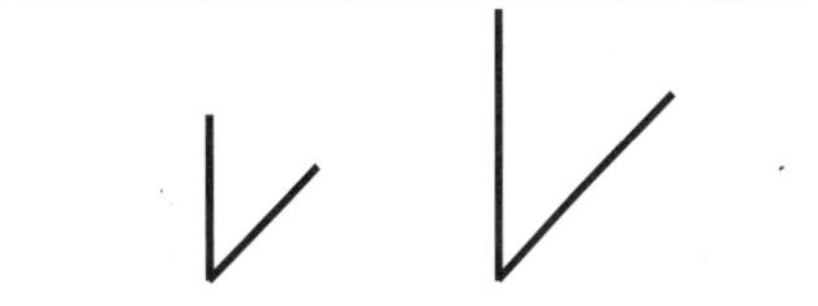

Discussion

Tasks 1-6, pp. 128-129

4: Point out that in a closed figure, we can't go from a point inside the figure to the outside without crossing a line. Have your student explain why A, B, and E are polygons, but the others are not. C is not a polygon because it has curved sides. D and F are not closed figures.

5: After counting the angles and sides, ask your student for a general rule about the number of angles compared to the number of sides for a polygon. They are equal.

2. **3** sides and **3** angles
3. 4
4. A, B, E
5. Quadrilaterals: 4 angles and 4 sides.
   Triangles: 3 angles and 3 sides
   Octagons: 8 angles and 8 sides
   Hexagons: 6 angles and 6 sides
   Pentagons: 5 angles and 5 sides
6.

P Pentagon	Q Hexagon	R Quadrilateral (Rectangle)	S Hexagon
T Octagon	U Triangle	V Octagon	W Pentagon

Activity

Ask your student for the names of some common quadrilaterals. Squares and rectangles are both quadrilaterals.

Poly-	many
-gon	angles
Penta-	5
Hexa-	6
Octa-	8

Point out that the shapes with a greater number of sides all have names that end with "-gon" similar to polygon. "Poly-" means "many", and "gon" means "angle" so polygon means "many angles". List the prefixes and write their meanings. You can tell your student that one way to remember the prefix "octa-" is to remember than an octopus has 8 arms.

Draw a convex polygon (all angles less than 180°, a straight line) and mark the angles with small curves. Tell your student that when we are asked to mark the angles of a polygon, it usually means to mark the angles on the inside of the polygon, even though the angle could also be marked on the outside.

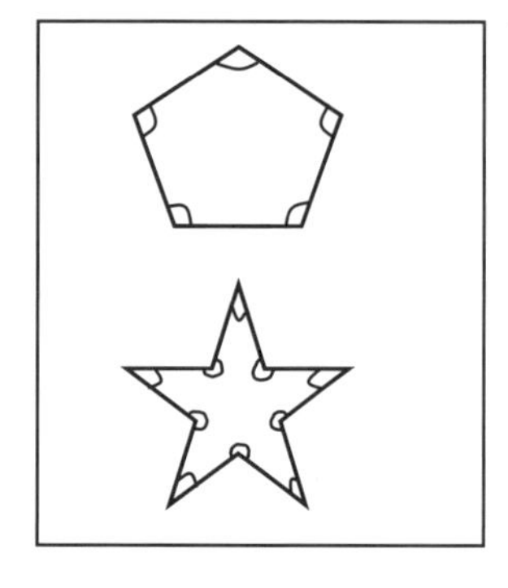

Draw a concave polygon (at least one angle greater than 180°, such as the star shown here) and ask your student to mark the angles. She should mark the interior angles, even ones that are greater than a straight line.

Workbook

Exercise 1, pp. 146-147 (Answers p. 141)

Reinforcement

Give your student a ruler and ask him to draw various types of polygons by name. For example, you could ask her to draw a triangle inside an octagon.

Extra Practice, Unit 12, Exercise 1, pp. 211-212

Test

Tests, Unit 12, 1A and 1B, pp. 227-234

Enrichment

If your student is interested, you can discuss the names of polygons with some other numbers of angles, some of which are shown in the table at the right. (The only names of polygons your student is expected to remember at this point are: triangle, quadrilateral, pentagon, hexagon, and octagon.) A triangle can be called a trigon, and a quadrilateral can be called a tetragon. Usually, after 6 angles or sides, polygons are just called n-gons.

Angles	Name
3	Triangle or Trigon
4	Quadrilateral or Tetragon
5	Pentagon
6	Hexagon
7	Heptagon
8	Octagon
9	Nonagon or Enneagon
10	Decagon
11	Hendecagon or Undecagon
12	Dodecagon
20	Icosagon
100	Hectogon
10000	Myriagon

Chapter 2 – Right Angles

Objectives

- Identify right angles.
- Classify angles as greater than, less than, or equal to a right angle.

Vocabulary

- Right angle

Notes

In this chapter, students will classify angles into three groups: right angle, less than a right angle, and greater than a right angle. They will then measure the angles in a polygon by comparing them to a right angle.

Measurement of angles in degrees will be learned in *Primary Mathematics* 4, as well as the terms acute and obtuse angles. You may want to introduce those terms now, but students are not required to learn them. An acute angle is less than a right angle and an obtuse angle is greater than a right angle.

When asked about the angles in a polygon, your student should assume that they should only consider the interior angles for now, even if the angle is reflex (greater than 180^0).

The activity in the lesson involves folding a piece of paper twice to create a right angle. This activity may seem unnecessary, because your student can simply use the corner of an index card to determine whether an angle is a right angle, but this activity does provide a concrete introduction to concepts the student will learn later. What this activity does is essentially divide the total distance around a point into fourths. (One fourth of 360^0 is 90^0, a right angle). In *Primary Mathematics* 4 students will be relating right angles to degrees and a fraction of the total degrees in a full turn around a point (a right angle is a fourth of a full turn). Make sure your student lines up the edges for the second fold.

Material

- Index card

(1) Identify right angles

Discussion

Concept p. 130

Have your student do this activity. She can use the angle formed to check the angles on the page and in the environment. Have her find out whether the corners on an index card are right angles. They are, so in later activities she can use the corner of an index card to compare angles to a right angle.

Angles b, c, d, and f are right angles

Tasks 1-3, p. 131

1: Ask your student what kind of polygon these two figures are. They are both quadrilaterals. Tell him that squares and rectangles are special types of quadrilaterals where all four angles are right angles.

3: You can also ask your student to name each type of polygon. Remind her that for questions like these, we are only looking at the inside angles of the polygon.

1. (a) 4 (b) 4
2. B has a right angle.
   C has an angle greater than a right angle.
3. P - 4 angles, 1 right angle. Quadrilateral
   Q - 5 angles, 2 right angles. Pentagon
   R - 4 angles, 2 right angles. Quadrilateral
   S - 5 angles, 3 right angles. Pentagon

Activity

Draw a figure similar to the one shown here and ask your student how many angles are less than a right angle, equal to a right angle, or greater than a right angle. There are three right angles, marked with little squares. The angle marked with a curve should be considered a greater than a right angle, since we are considering only interior angles.

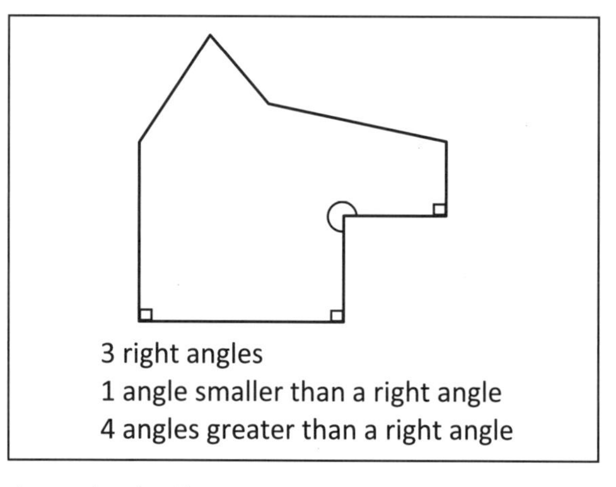

3 right angles
1 angle smaller than a right angle
4 angles greater than a right angle

Ask your student if a quadrilateral could have 0 right angles (yes) and then have him sketch some examples. Then, ask him if a quadrilateral could have only 3 right angles. Let him experiment to see if he can draw such a quadrilateral. Ask him if a triangle could have 2 right angles, and if a pentagon could have 4 right angles. (They cannot.)

Workbook

Exercise 2, pp. 148-150 (Answers p. 141)

Reinforcement

Extra Practice, Unit 12, Exercise 2, pp. 213-214

Test

Tests, Unit 12, 1A and 1B, pp. 235-242

Chapter 3 – Quadrilaterals and Triangles

Objectives

- Identify parallel and intersecting lines.
- Identify parallelograms.
- Recognize rectangles, squares, and rhombuses as types of parallelograms.
- Identify various attributes of different types of parallelograms.
- Identify equilateral, isosceles, scalene, and right triangles.

Vocabulary

- Intersecting lines
- Parallel lines
- Parallelogram
- Rhombus
- Equilateral triangle
- Isosceles triangle
- Scalene triangle
- Right triangle

Notes

In this chapter, students will identify intersecting lines and classify different types of parallelograms and triangles.

Intersecting lines are lines that cross each other. The point where they cross is called the point of intersection. Parallel lines are lines that will never cross each other no matter how far they are extended. The distance between the two lines is always the same. At this level, students can identify parallel lines by appearance; if it looks parallel, it is. There will not be lines that are almost parallel but not quite.

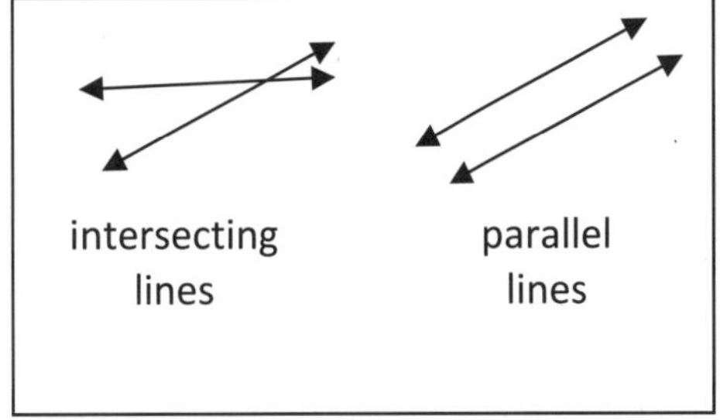

A parallelogram is a quadrilateral where both pairs of opposite sides are equal to each other. If a parallelogram has 4 equal sides, it is a rhombus. If the parallelogram has 4 right angles, it is called a rectangle. A square is both a rectangle and a rhombus.

Parallelogram: Opposite sides are parallel and equal.

Rectangle: Parallelogram with 4 right angles.

Rhombus: Parallelogram with 4 equal sides.

Square: Parallelogram, Rhombus, and Rectangle

In the US, a quadrilateral where only one pair of sides is parallel and the other is not is called a trapezoid. In other countries, this is called a trapezium. Students will learn about trapezoids in *Primary Mathematics* 4A.

Students will classify triangles based on their sides. An isosceles triangle has two equal sides. An equilateral triangle has three equal sides. An equilateral triangle is also considered to be isosceles in this curriculum. A scalene triangle has no equal sides.

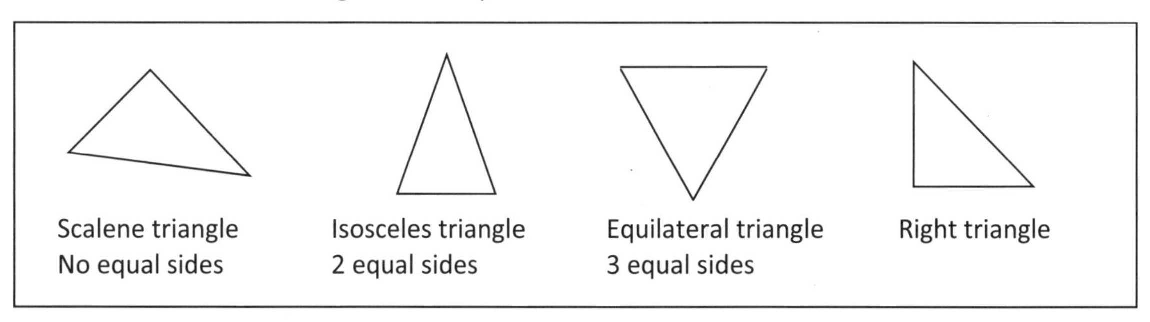

Students will also identify a triangle with a right angle as a "right triangle." It can also be called a "right angled triangle." A right triangle can be isosceles or scalene.

In *Primary Mathematics* 4, students will learn to classify other triangles based on their angles. A right triangle has a right angle, an acute triangle has all angles less than a right angle, and an obtuse triangle has one angle greater than a right angle.

(1) Identify quadrilaterals and triangles

Discussion

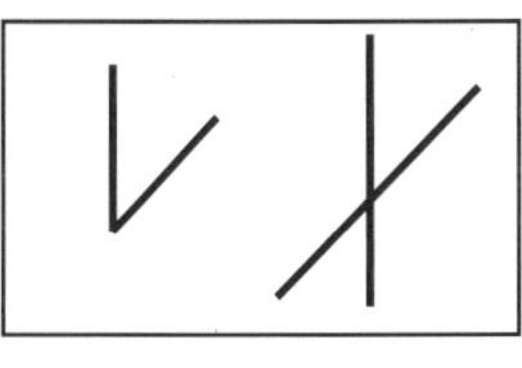

Draw an angle. Remind your student that an angle is formed where two lines come together, or **intersect**. There are actually 2 angles, one "inside" and one "outside." Extend the lines beyond the vertex. When two lines cross each other, 4 angles are formed. Lines that cross each other are called **intersecting lines**.

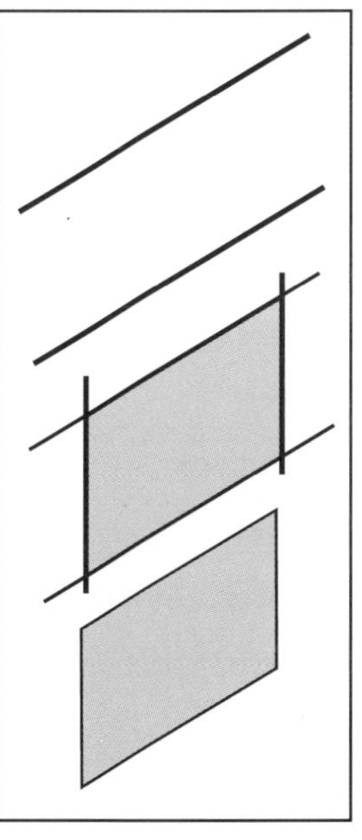

Draw two parallel lines. (An easy way to do this is to draw two lines on either side of a ruler.) Tell your student that lines that will never cross each other no matter how long they are extended are called **parallel lines**. The distance between the lines stays the same. Ask your student for some examples in the environment of parallel lines, such as railroad tracks.

Now draw two more parallel lines between the two you already have. (One way to do this is to line up the wide side of the ruler with the bottom of the page, draw a line between the two lines that is perpendicular to the bottom of the page, then slide the ruler along to draw the other side.) Tell your student that the new lines are also parallel lines. The shape that the two sets of parallel lines form is called a **parallelogram**. Ask her what else such a shape could be called. Since there are 4 sides, it is a quadrilateral. Since the two sets of lines are parallel and so the same distance apart all along their length, what can we say about the length of the two lines in each set of parallel sides? They are equal to each other.

Discussion

Concept p. 132

A, C, and F are parallelograms

Tasks 1-4, p. 133

You can ask your student to explain his answers for tasks 1-3.

1: Both sets of opposite sides are parallel.

2: Both sets of opposite sides are parallel and all four sides are equal, so a square is both a parallelogram and a rhombus.

3: Both sets of opposite sides are parallel, but all four sides are not equal. So a rectangle is a parallelogram but not a rhombus.

1. Yes
2. Yes
   Yes
3. Yes
   No
4. (a) A, B, C, D
   (b) A, B, C, D
   (c) A, B

Ask your student what makes the square and the rectangle different from other parallelograms. They have 4 right angles.

Make sure your student understands that a rhombus is a special type of parallelogram; it has 4 equal sides. All rhombuses are parallelograms, but not all parallelograms are rhombuses. A rectangle is also a special type of parallelogram; it has 4 right angles. All rectangles are parallelograms but not all parallelograms are rectangles. A square is a type of rectangle that is also a rhombus.

Tasks 5-8, p. 134

5. No

6. (a) scalene
   (b) isosceles
   (c) equilateral

7. (a) yes
   (b) yes
   (c) no

8. (a) D
   (b) A, D
   (c) B, C
   (d) B

6: Discuss the three triangles shown here with regard to the length of their sides. Make sure your student understands that an equilateral triangle has 3 equal sides, an isosceles triangle has 2 equal sides, and a scalene triangle has 0 equal sides. Tell your student that an equilateral triangle is also an isosceles triangle, since it also has 2 equal sides. So an equilateral triangle is a special type of isosceles triangle where the third side is also equal to the other two.

7: To show that a right triangle cannot be equilateral, you can draw the two sides, and then draw the third side farther and farther out at measured intervals and measure the third side. In all cases, the third side will be longer than the other two.

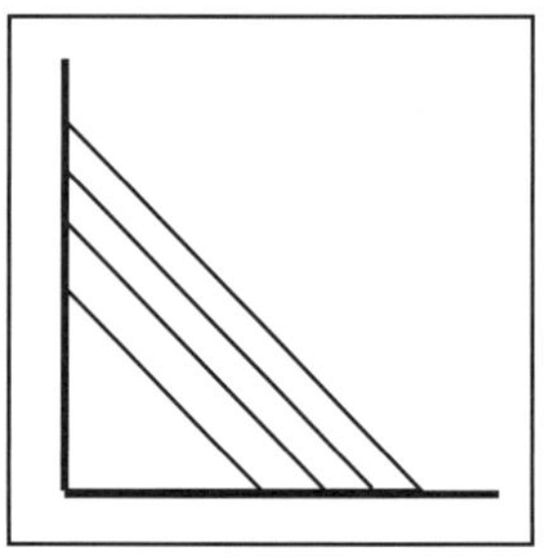

Workbook

Exercise 3, pp. 151-152 (Answers p. 142)

Reinforcement

Extra Practice, Unit 12, Exercise 3, pp. 215-216

Test

Tests, Unit 12, 3A and 3B, pp. 243-250

Enrichment

You may want to discuss the origin of the names for triangles and the meanings of the prefixes and suffixes. For some students, this can help them remember which name describes which kind of triangle.

equi-: equal   lateral: sides
Equilateral triangle: triangle with equal sides

Iso-: equal   -skelos: legs
Isosceles triangle: Triangle with "equal legs"

Scalenus: unequal
Scalene triangle: triangle with unequal sides.

Chapter 4 – Solid Figures

Objectives

- Identify cubes, prisms, pyramids, cylinders, cones, and spheres.
- Classify and sort common solids by faces and by number of edges and vertices.
- Identify common solid shapes that make up a more complex object.

Vocabulary

- Cube
- Prism
- Rectangular Prism
- Triangular Prism
- Pyramid
- Cylinder
- Cone
- Sphere
- Face
- Curved Surface
- Edge
- Vertex
- Base
- Apex

Notes

In *Primary Mathematics* 2B, students learned to identify some basic solids: cubes, prisms, rectangular prisms, pyramids, cones, and spheres. They counted the number of faces, edges, and vertices on these shapes. This chapter is almost entirely a review. The only new material added is distinguishing between a triangular pyramid and a rectangular pyramid, and naming the solids in an object where the shapes are combined. Students will only examine right prisms where the top and bottom polygons lie on top of each other so that the rest of the faces are not only parallelograms, but also rectangles.

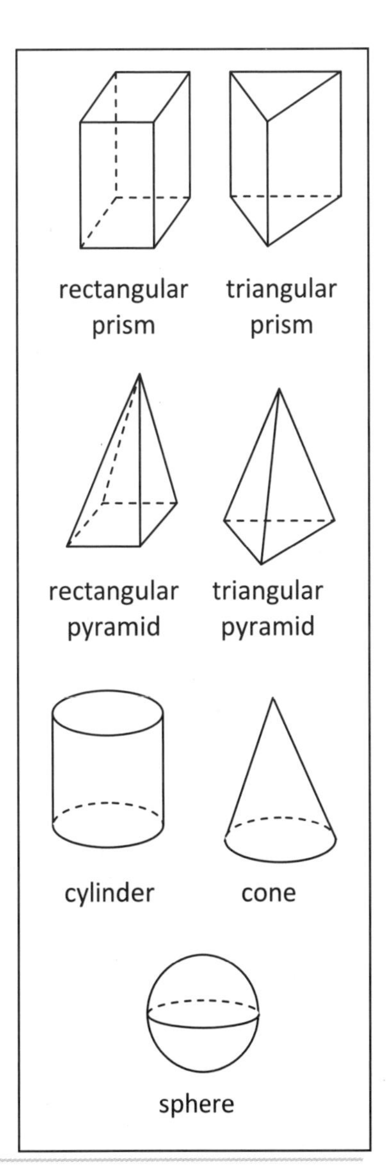

In *Primary Mathematics* 2B, students identified faces and edges that were curved or flat. In *Primary Mathematics* 4, students will be told that a *face* is a flat surface and an *edge* is formed where two faces meet. At the primary level, the term *face* will not be restricted to a polygon, as it will be later in a formal study of geometry, and so an *edge* can be curved. At this level, you should restrict the term *face* to a *flat* surface. A *surface* can be curved or flat. A sphere has one curved surface, but no faces. A cylinder has one curved surface and two flat surfaces, or two faces. A cone has one curved surface and one face.

A *vertex* is where three or more edges meet. The term vertex can also be used for the *apex* of a cone.

Material

- Models of cubes, prisms, rectangular prisms, pyramids, cylinders, cones, and spheres

(1) Identify solids

Discussion

Concept pp. 135-136

> A square pyramid (rectangular pyramid) has 8 edges.
> A square pyramid has 5 vertices.
> A sphere has no faces.

Use actual solids that your student can touch and move. Have your student name the solids, count the number of faces or curved surfaces, edges, and vertices on each.

Ask your student for similarities between the rectangular prism and triangular prism to help them determine why they are both "prisms." A prism has two polygonal faces that are the exact same shape and size, which we can call bases. The remaining faces are all parallelograms. If the bases are directly on top of each other, as in the examples in the text, the remaining faces are rectangles. The prism is named by the shape of the bases; a triangular prism has triangles for the two opposite bases, and a rectangular prism has rectangles (which could be squares) for the two opposite bases. Make sure your student understands that a cube is a special type of rectangular prism where all the faces are squares.

Ask your student for similarities between the two pyramids and how they differ from prisms. Pyramids have a polygon for one base, and all the other faces are triangles that meet at a common polygon vertex, or *apex*. They are named according to the shape of the base.

You can ask your student to identify these solids in the environment. Your student can also look for prisms or pyramids with bases other than rectangles or triangles.

Tell your student that for mathematics we will define a *face* to be flat. So a cylinder has 3 surfaces, 1 curved surface and 2 flat faces. A cone has 2 surfaces, one curved and 1 flat face. A sphere has 0 faces, but 1 curved surface.

Task 1, p. 136

> 1. A: cone, rectangular prism
>   B: pyramid (can't tell which kind from the drawing), cube or rectangular prism.
>   C: sphere, cylinder

Most objects in the environment are made up of a variety of solid shapes. The ones shown here in the textbook are somewhat lame. You can have your student identify common shapes in more complex shapes in the environment. For example, a house could contain the shape of a rectangular prism and a triangular prism which includes the roof. A bird feeder could contain three cylinders between the top and the bottom, or perhaps contain an octagonal prism.

Workbook

Exercise 4, pp. 153-154 (Answers p. 142)

Reinforcement

Extra Practice, Unit 12, Exercise 4, pp. 217-218

Test

Tests, Unit 12, 4A and 4B, pp. 251-158

Note: Test A, for 5 p. 252: Either change the problem to "How may surfaces does a sphere have?" or change the answer to 0. For problem 9, the number of faces for a cylinder should be 2, and for a cone should be 1. Or, change the heading to "Number of surfaces."

Enrichment

Topology is a field of mathematics that is like geometry in that it deals with points and lines, but unlike regular geometry it allows objects to change shape and size. The Möbius strip is an interesting introduction to topology.

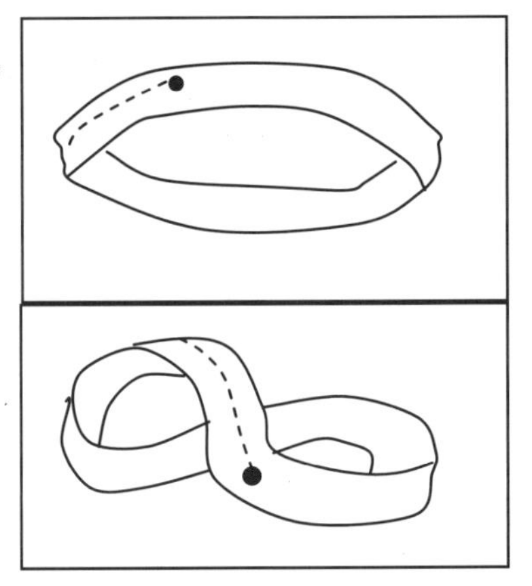

Take a strip of paper and tape it in such a way that it forms a circle. Have your student mark a point on one side and then draw a line the length of the strip along the middle of the strip. When he returns to the point, he will have drawn a line on one side of the paper. This strip of paper has two sides.

Take another strip of paper and make a circle but put a half twist in it before taping it. This is a Möbius strip. Have your student mark a point and draw a line along the middle of the strip. Ask her what happens. The line will pass the marked point on the other side of the paper, and to get back to the side marked will apparently go another circuit. So without lifting the pencil, there will be no unmarked side of the strip. So a Möbius strip has only one side.

Make a mark at the very edge of the strip. Have your student run his finger along the edge and tell you how many edges the strip has. A Möbius strip has only one edge.

Have your student cut both strips in half along the lines she has drawn and see what happens.

Your student can then investigate what happens if the strip of paper is twisted a whole turn, or a turn and a half, before taping.

After cutting the Möbius strip in half along the line, your student can find out what happens if she cuts it again along its length.

Your student can see if twisting the strip twice or three times before taping gives a one-sided or two-sided surface.

Review 12

Review

Review 12, pp. 137-138

Workbook

Review 12 pp. 155-158 (Answers p. 143)

Test

Tests, Units 1-12, Cumulative A and B, pp. 259-272

1. (a) $\frac{1}{2}, \frac{5}{8}, \frac{3}{4}$ (b) $\frac{3}{10}, \frac{1}{2}, \frac{3}{5}$

2. 9 cm x 3 = 27 cm.
   The total length of the sides is **27 cm**.

3. (a) 3 (b) 7
   (c) 4700 (d) 40
   (e) $\frac{4}{10}$ (or $\frac{2}{5}$) (f) $\frac{2}{6}$ (or $\frac{1}{3}$)

4. 30 cm x 8 = 240 cm = 2 m 40 cm.
   The wire was **2 m 40 cm** long.

5. Men 1360 | Women ? | Children 240
   2500

   (a) 1360 + 240 = 1600
   2500 – 1600 = 900
   There were **900** women.
   (b) 1360 + 900 = 2260
   2260 – 240 = 2020
   There were **2020** more adults than children.

6. Cookies left: 286 – 30 = 256
   Make groups of 8: 256 ÷ 8 = 32
   Each of the 32 groups of 8 was sold for $1.
   She received **$32**.

7. (a) A hexagon has 6 sides and 6 angles.
   (b) An equilateral triangle has 3 equal sides.
   (c) A rectangle has 4 right angles.

8. $1 - \frac{1}{5} - \frac{1}{5} = \frac{3}{5}$
   John got $\frac{3}{5}$ of the pizza.

9. 1 km – 580 m = 420 m
   Tom's house is **420 m** from the library.

10. (a) The total weight is **2 kg 400 g**.
    (b) Weight of butter: 300 g x 3 = 900 g
    Weight of flour:
    2 kg 400 g – 900 g = 1 kg 500 g
    The flour weighed **1 kg 500g**.

11. (a) A
    (b) B, C

Workbook

Exercise 1, pp. 146-147

1. (a) ☐ (b) ☐ (c) ☐
   (d) ☑ (e) ☐ (f) ☑

2.

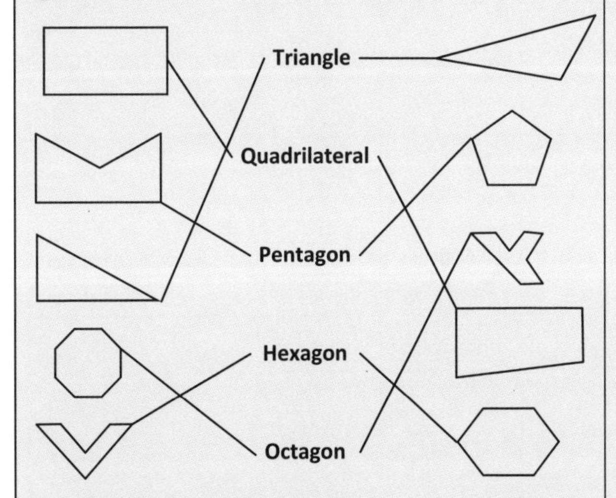

3.

Figure	Number of sides	Number of angles	Name of polygon
A	4	4	Quadrilateral
B	5	5	Pentagon
C	3	3	Triangle
D	4	4	Quadrilateral
E	6	6	Hexagon
F	5	5	Pentagon

Exercise 2, pp. 148-150

1.

Angle	Smaller than a right angle	Bigger than a right angle	Equal to a right angle
a	✓		
b			✓
c		✓	
d		✓	
e			✓

2.

3.

Figure	Number of sides	Number of angles	Number of right angles
A	4	4	2
B	4	4	2
C	4	4	4
D	4	4	2
E	4	4	1
F	5	5	3
G	3	3	1
H	4	4	4
I	4	4	2

Workbook

Exercise 3, pp. 151-152

1.

Parallelogram	Rhombus	Rectangle	Square
A, B, C, D, E, G	A, B, D, G	A, C, D	A, D

2.

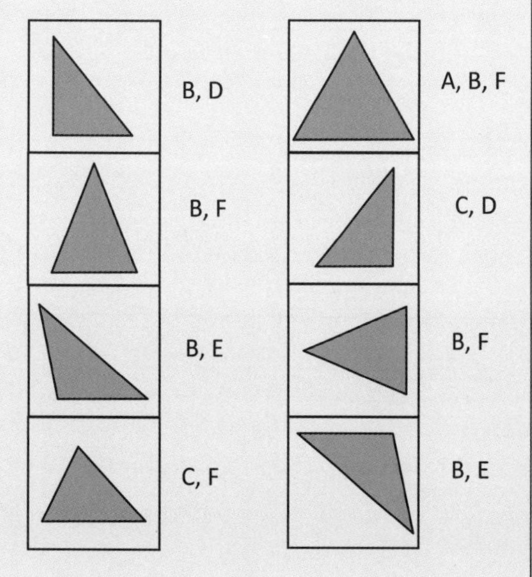

Exercise 4, pp. 153-154

1.

Shape	Faces	Edges	Vertices
Cube	6	12	8
Rectangular prism	6	12	8
Triangular prism	5	9	6
Triangular pyramid	4	6	4
Rectangular pyramid	5	8	5
Cylinder	2	2	0
Cone	1	1	1
Sphere	0	0	0

2. A, B, F
   C, D
   B, E
   B, E, F

Workbook

Review 12, pp. 155-158

1. (a) 32, 40, 48
   (b) 63, 54, 45
   (c) 175, 195, 215
   (d) 1934, 1734, 1534

2. (a) 6 + 14 + 10 + 4 = 34
   34 students wear glasses.
   (b)

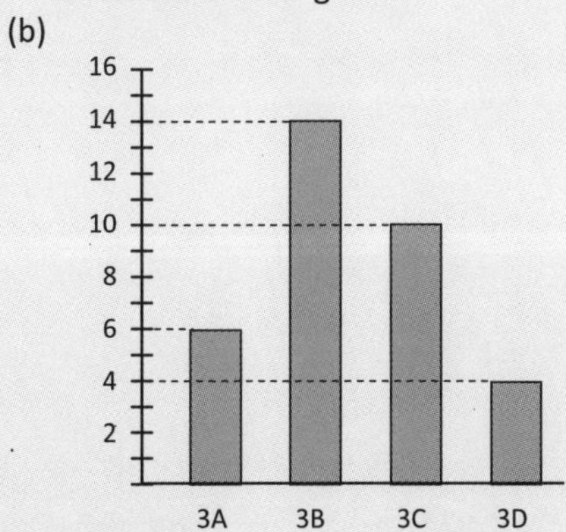

3. (a) 3 h 50 min
   (b) 11:35 a.m.

4. (a) 6258 (b) 323 R 2 (c) \$7.50 (d) \$0.55

5. (a) $\frac{3}{4}$ (b) $\frac{5}{12}$ (c) $\frac{2}{5}$ (d) 1

6. (a) $\frac{1}{4}$ (b) $\frac{1}{100}$ (c) $\frac{3}{4}$ (d) $\frac{1}{2}$

7. (a) $\frac{4}{7} > \frac{3}{7}$ (b) $\frac{3}{5} > \frac{3}{8}$
   (c) $\frac{1}{2} > \frac{3}{9}$ (d) 5 = 5

8. 6

9. 5

10. b

11.

4 oz 10 oz
Apple
Banana
Grapes

Weight of bananas: 4 oz + 10 oz = 14 oz
Weight of grapes: 14 oz x 3 = 42 oz
42 oz = 32 oz + 10 oz = 2 lb + 10 oz
The grapes weigh **2 lb 10 oz**.

12. 6 gal – 3 gal 1 qt = 2 gal 3 qt
    2 gal 3 qt of milk was left.

13. (a) The spinner lands on A, B, or C.
    (b) A

Unit 13 – Area, Perimeter and Volume

Chapter 1 – Area

Objectives

- Measure and compare area in square units.
- Measure and compare area in square centimeters.
- Measure and compare area in square inches.

Vocabulary

- Area
- Square unit
- Square centimeter
- Square inch

Notes

In this chapter, students will be introduced to the concept of area. They will measure and compare areas first in non-standard square units and then in square centimeters and square inches.

The area of a surface is the amount of material needed to "cover" it completely. At this level, we will only deal with flat surfaces. Area is measured in square units. If the area of a figure is 4 square centimeters, then it covers the same amount of flat space as would four squares with 1 cm sides.

Material

- Flat square tiles or square pieces of cardboard or paper
- Centimeter graph paper (Appendix p. a14, 4-5 copies)
- Inch graph paper (Appendix p. a15, 2-3 copies)
- Optional: Paper sheets 1 foot square
- Optional: 4 meter sticks or string or cardboard strips 1 meter long

(1) Measure area in square units

Activity

Tell your student that area is the amount of flat space covered. Discuss the term as it is used in everyday use, such as area rugs, the area of a room, the number of acres of land, etc.

Use some square tiles or square pieces of paper. You can copy the inch graph paper in the appendix and cut out the squares, or use pattern block squares if you have them. Tell your student that area is measured in square units. We have to define the size of the square unit, and then we can give the area as the number of square units that fit in that area. Put several squares together and have him count them to find the area in square units. Cut some of the squares in half diagonally (If you have pattern blocks use the triangle and show that two triangles fit on the square). Ask him for the area of each resulting triangle (half a square unit). Have him create a shape with squares and half squares and then tell you the area in square units. Then have him move the squares and half-squares around to make another shape with the same area. Tell him both shapes have the same area; they cover the same amount of flat space. So we can say that they have the *same area* or they are the same size. Add another square unit to the figure and ask your student for the area. Tell your student that this new figure is bigger; it covers a larger area.

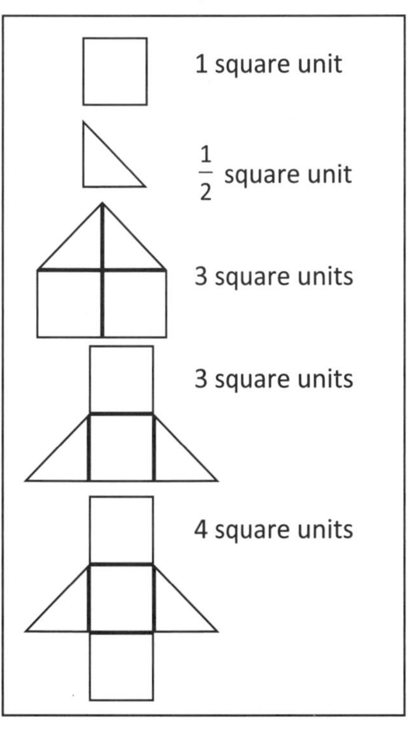

Discussion

Concept p. 139

Emphasize that all these figures are the same size even though they are not the same shape. They all have the same area

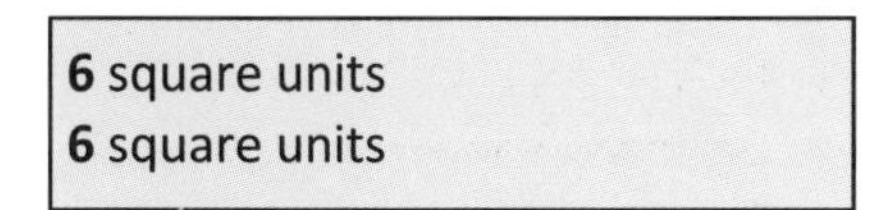
6 square units
6 square units

Tasks 1-2, pp. 140-141

1: Ask your student which of these shapes is biggest. The pink rectangle is, since it area is greatest. Point out that we can only compare the areas of the figures to each other if the square units are the same.

2: See if your student can find the area of S without help. If she needs help, cut out two squares as shown here, draw a diagonal line, color one half, and then cut it in half as shown here to show that the shaded part has the area of 1 square unit. Ask your student which of the shapes is the biggest shape and which is the smallest shape.

1. (a) 24
   (b) 12
   (c) 22

2. P: 7 square units
   Q: 8 square units
   R: 8 square units
   S: 12 square units

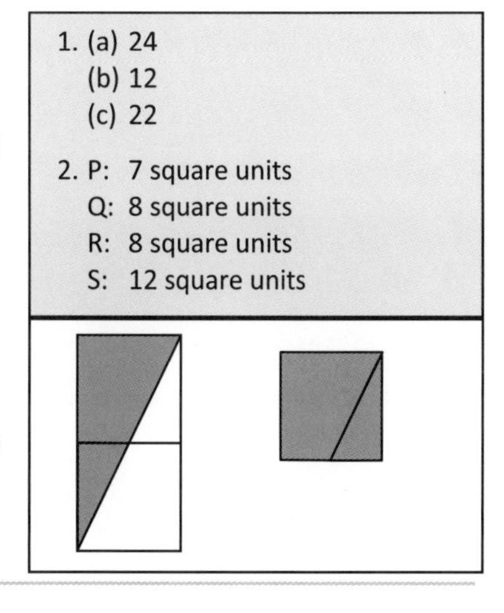

Workbook

Exercise 1 pp. 159-162 (Answers p. 161)

Enrichment

Use a copy of the centimeter graph paper. In two of the squares draw an irregular curved line through them. Shade one part of each to make a shape. Ask your student to estimate the shaded area in each square; about a half, more than a half, or less than a half. If the shaded area in one is a more than a half, and in the other less than a half, we can estimate the total shaded area to be *about* 1 square unit.

Draw an irregular shape on square graph paper and shade it. Ask your student to count the whole squares and write the total down. Then have her pair and mark off squares that are less than half shaded with squares that are more than half shaded and thus estimate the total area of the shaded figure.

Give your student some square graph paper. Have him trace his hand and find the area of his hand prints. He can compare the areas of his hand with the area of another person's hand, or with her foot print.

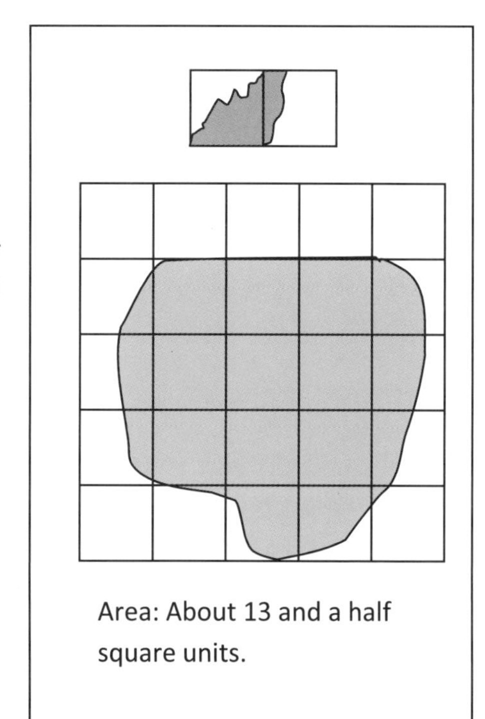

(2) Compare areas

Discussion

Tasks 3-6, pp. 141-143

6: See if your student can estimate the number of squares that will cover the figures. Then you can have your student use centimeter squares or cubes to cover the figures.

3. A: **7** square units
   B: **6** square units
   C: **8** square units
   C is biggest
   B is smallest
4. A: 6 square units B: 5 square units
   C: 13 square units D: 6 square units
   E: 7 square units F: 10 square units
   G: 10 square units H: 12 square units
   B has the smallest area.
   C has the greatest area.
5. P: 7 square units Q: 6 square units
   R: 7 square units S: 5 square units
   P and **R** are the same size.
6. (a) 15 (b) 9

Workbook

Exercise 2, pp. 163-166 (Answers p. 161)

Enrichment

Draw 4 squares in a larger square as shown here, or outline or shade 4 squares on square graph paper. Ask:

⇒ How many squares of 1 square unit are there? (4)

⇒ Is there a larger square? (yes)

⇒ What is the area of the larger square? (4 square units)

⇒ How many total squares does the figure have? (5)

Draw or shade 9 squares in a larger square. Ask:

⇒ How many squares of 1 unit square? (9)

⇒ How many squares of 4 unit squares? (4, they overlap)

⇒ Is there an even larger square? (Yes)

⇒ What is the area of the larger square? (9 square units)

⇒ How many total squares does the figure have? (14)

Draw 16 squares. Ask:

⇒ How many total squares does the figure have? (30)

Help your student determine how many there are of each size. There are 16 1-unit squares, 9 4-unit squares, 4 9-unit squares, and 1 16-unit square.

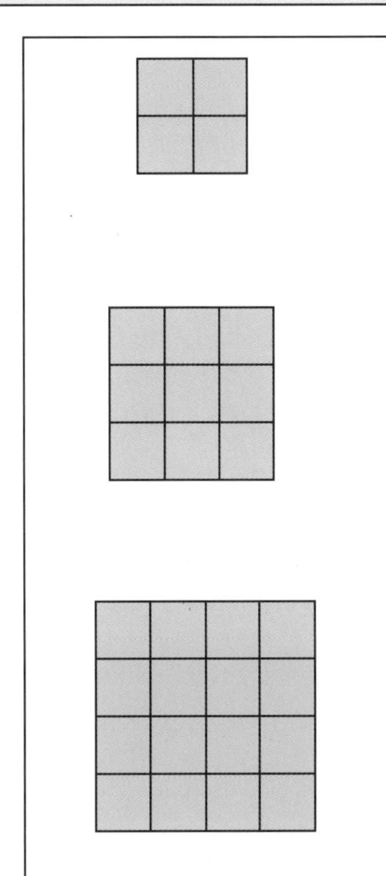

(3) Measure area in square centimeters or square inches

Discussion

Tasks 7-11, pp. 144-146

Before doing task 7, tell your student that, just as with other types of measurement, we need to have a standard way of measuring area. Ask your student for some units of length. Possible answers are centimeter, meter, yard, inch, feet, etc. Tell your student that area is measured in squares with standard lengths for the sides.

7: You can have your student measure the sides of the square with a ruler to see that the picture in the text is the exact size of a square centimeter. Point out that a 2-cm square is a square that is 2 cm long on each side.

8: You can have your student use a copy of the square centimeter page in the appendix to solve these.

10: Ask your student which shape has the greatest area (B), which shape has the smallest area (H), and which shapes have the same area (A and C, D and F, E and G)

11: You can have your student measure the sides of the square inch in the text.

7. 4 square cm
   9 square cm
   16 square cm
8. 25 square cm
   100 square cm
9. 10 square cm
10. A: 5 square cm
    B: 8 square cm
    C: 5 square cm
    D: 6 square cm
    E: 7 square cm
    F: 6 square cm
    G: 7 square cm
    H: 4 square cm
11. A: 6 square inches
    B: 4 square inches
    A has the greater area.
    B has the smaller area.

Workbook

Exercise 3, pp. 167-169 (Answers p. 161)

Reinforcement

Extra Practice, Unit 13, Exercise 1, pp. 223-228

Test

Tests, Unit 13, Chapter 1, A and B, pp. 273-283

Enrichment

Discuss some other units of area. Tape some sheets of paper together to make square feet. Make 9 of them. Show one to your student and have him measure the side. Ask him for the unit of area. (1 square foot. The picture shown at the right is scaled down and just given to illustrate the next point.)

Arrange the 9 square feet on the floor in a 3 x 3 arrangement. Ask your student if he can give the unit of area for this figure. If necessary, remind him that there are 3 feet in a yard, so one side of the figure is one yard long. (1 square yard.) Ask him how many square feet are in a square yard. Point out that even though 3 feet make a yard, this does not mean that 3 square feet make a square yard.

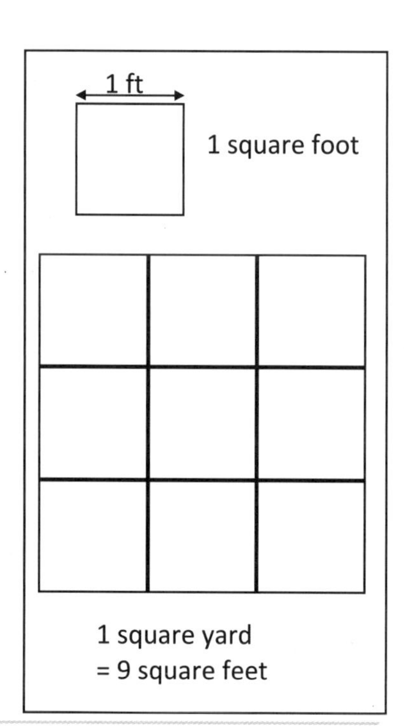

Use 4 meter sticks or 4 pieces of string a meter long to form a square one meter on its side, or draw a square meter on the sidewalk with chalk. Tell your student that this is another unit of measurement, a square meter.

Put a square centimeter in the corner of the square meter. Note the relative sizes. Ask your student how many centimeters there are in a meter. There are 100 cm in a meter. So the square meter is 100 cm on the side. Tell your student it is made up of 10,000 square centimeters.

You may want to discuss some other units of area. In the metric system, a square kilometer is a square with 1000 meters on each side. An *are* is a square with 10 meters on each side. A *hectare* is a square with 100 meters on each side.

In U.S. customary measurement, an *acre* can be covered up with 4840 square yards, and a square mile can be covered up with 640 acres.

Discuss scale drawings. In a map, each square mile or square kilometer is drawn smaller so that it can fit on the map. Show your student the scale on some maps. The actual area of the land represented by the map is much bigger.

Chapter 2 – Perimeter

Objectives

- Understand perimeter.
- Measure the perimeter of figures.
- Compare the area of a rectilinear figure to its perimeter.
- Understand that figures can have the same area but different perimeters.
- Find the perimeter of polygons when given the length of the sides.

Vocabulary

- Perimeter

Notes

In this chapter students will learn that the perimeter of a figure is the distance around the outside of the figure.

To find the perimeter of a rectilinear figure drawn with squares, we can count the units of length along the outside. Your student will need to be careful to count each unit length only once, and not to count a square corner as only one unit. The figure at the right has a perimeter of 12 units.

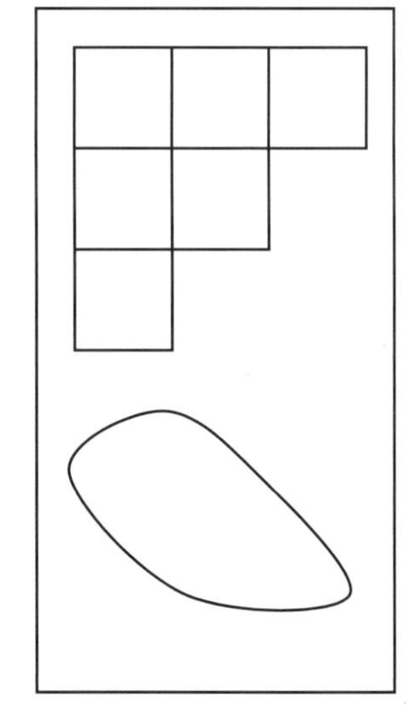

Students will use a string to measure the perimeter of figures with curved sides.

Different shapes may have the same area but different perimeters, or the same perimeter but different areas, or two shapes could have the same perimeter and area but be different shapes.

If the length of the sides is measured or given, we find the perimeter by adding up the lengths of the sides.

The unit of measurement can be any of the standard units. Make sure the students understand that the figures in the textbook can be scaled down.

Students will find the perimeter of a polygon when the lengths of the sides are given by adding the lengths. Formulas for finding the perimeter of squares and rectangles will be taught in *Primary Mathematics* 4. Your student may derive some formulas, for example the perimeter of a square is four times the length of a side, but it is not necessary to formally teach formulas for perimeter at this level.

Material

- Piece of wire cut to a whole number of centimeters, or three straws each cut to a whole number of centimeters
- Centimeter graph paper (Appendix p. a14)
- Appendix p. a16
- String
- Ruler

(1) Measure perimeter

Activity

If you have a straight piece of wire, show it to your student and ask him to measure its length. Then bend it into a triangle and ask him for the distance around the triangle. Or, give your student three straws of different lengths, ask your student to find the total length, form a triangle with the straws, and then ask him for the distance around the triangle.

Tell your student that we call the distance around a figure the perimeter. You may want to explain that "perimeter" comes from Greek: *peri* means "distance around" and *metreo* means "to measure". When we measure the distance around a figure, we are finding its perimeter. You can also point out that there is the word RIM in perimeter; we are finding the distance around the "rim" of a figure.

Perimeter: Distance around

Peri: distance around
Metreo: to measure

Discussion

Concept p. 147

Use appendix p. a16 Instead of the page in the textbook; the squares in the first printing are not quite square centimeters as they should be.

Perimeter of each should be 24 cm.

Ask your student to find the perimeter around the rectangle and square first by counting the sides of the squares. Remind her that perimeter is the distance around the figure, as if a very small bug were following the edge. Make sure your student counts the two centimeters around a corner square. Then have your student find the perimeter around the triangle, using a ruler to measure the hypotenuse (its length should be 10 cm).

Tasks 1-4 pp. 148-149

1: Using string or thread to measure perimeter emphasizes that perimeter is the distance around a shape.

3: Figures with the same area can have different perimeters.

4: Figures with the same perimeter can have different areas.

1. Y

3. (a) 6 square centimeters
(b) A - 12 cm
B - 14 cm

4.

Figure	Area Square cm	Perimeter cm
A	8	12
B	5	12
C	7	12

(a) no
(b) yes

Practice

Task 5, p. 149

5.

Figure	Area Square cm	Perimeter cm
P	7	14
Q	8	18
R	9	16
S	8	16
T	7	14

(a) Q and S
(b) R and S
(c) P and T

Activity

Use linking cubes, square cut-outs, the squares from pattern blocks, or centimeter graph paper to investigate the relationship of area to perimeter.

Ask your student what she could look for in figures with the same areas, such as those in task 3, to know if they are going to have different perimeters. Allow her to investigate by making various shapes with areas of 6 square units and then finding the perimeters. The sides must touch all along their entire length. Ask her if she notices anything about the shapes that might predict whether one has a greater or smaller perimeter than the other.

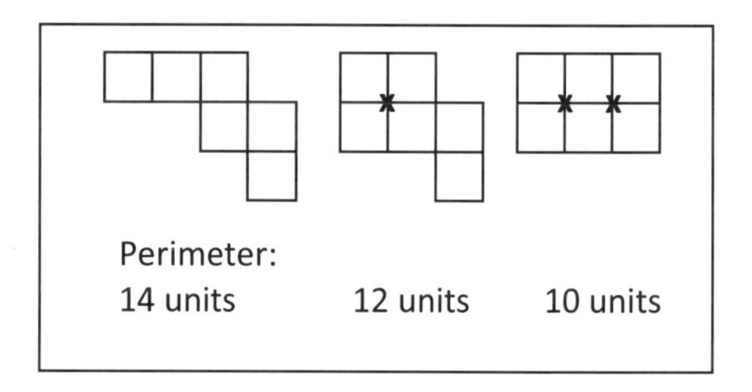

Any figure where a corner has squares on all four sides will have a smaller perimeter than a figure in which no corner has squares on all four sides, and the figure with most such corners has the smallest perimeter.

Some students have the misconception that if the area of one figure is greater than the area of another, then the perimeter must also be greater. Have your student investigate whether the perimeter necessarily increases if the area increases. It does not.

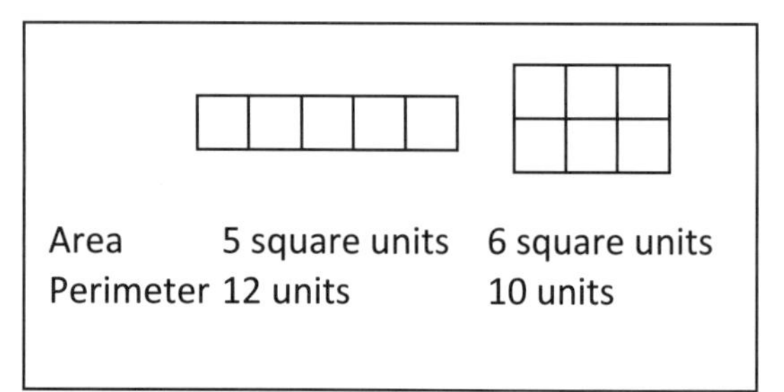

(2) Find the perimeter given the length of the sides

Activity

Draw a square made up of cm squares, such as a 3 by 3 square, and label the sides 3 cm. Use arrows to show that 3 cm is the length of the entire side, not the length of the side of one of the smaller squares. The squares you draw do not have to be exactly 3 cm squares. Tell your student that figures on paper may be scaled down to fit, like on a map. So that rather than having a student measure the sides, the length of the sides are labeled. Ask your student for the perimeter. She can count the sides of the square.

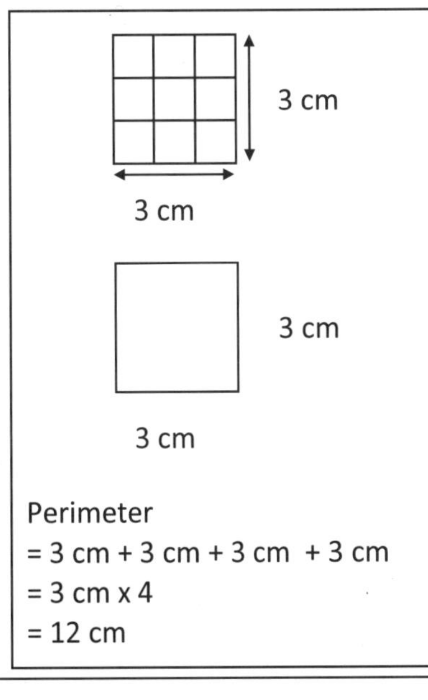

Erase the inside lines and the arrows. Tell your student that the labels are giving the length of the sides and ask her if we can find the perimeter without seeing each square. The perimeter is the sum of the lengths of the sides. Since all the sides are equal, the perimeter can be found by multiplying a side by 4.

Draw a rectangle and label two of the adjacent sides. Ask your student to find the perimeter. He should realize the opposite sides have the same length. Your student can simply add the length of the sides to find the perimeter. Since there are two pairs of equal sides, your student may realize that we can add the length and the width and then multiply by two to find the perimeter.

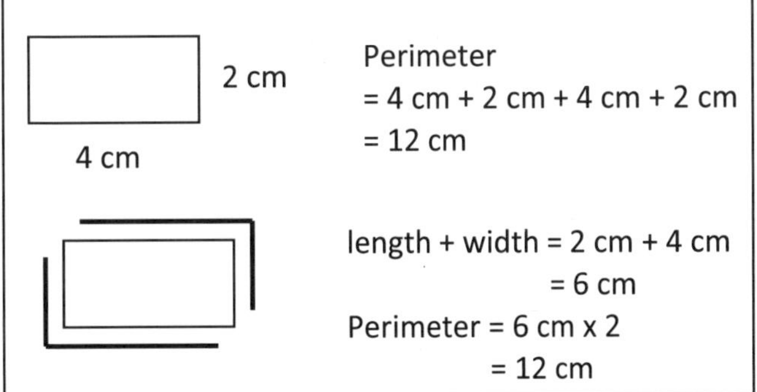

Practice

Tasks 6-7, p. 150

6. (a) 24 cm
   (b) 32 cm

7. A: 25 cm B: 34 in.
   C: 30 cm D: 39 in.

Workbook

Exercise 4, pp. 170-172 (Answers p. 162)

Reinforcement

Extra Practice, Unit 13, Exercise 2, pp. 229-232

Test

Tests, Unit 13, 2A and 2B, pp. 285-283

(Note: Problem 13 on Test A is challenging.)

Chapter 3 – Volume

Objectives

- Interpret 2-dimensional drawings of unit cubes.
- Build solids from models drawn on isometric dot paper.
- Determine the number of cubes used to make a solid.
- Find the volume in cubic units of 2-dimensional representations of solids made of unit cubes.

Vocabulary

- Volume
- Cubic units

Notes

In this chapter, students will learn that volume is the amount of space a solid occupies, and that volume is measured in cubic units. They will find the volume of a solid made up of unit cubes drawn in 2 dimensions, such as those shown below. They need to be able to interpret the solid; that is, visualize it as a 3-dimensional object, which includes cubes they can't see in the 2-dimensional drawing. In order to do this, they will first explore solids drawn in 2 dimensions with the aid of isometric dot paper and build solids to match the drawings.

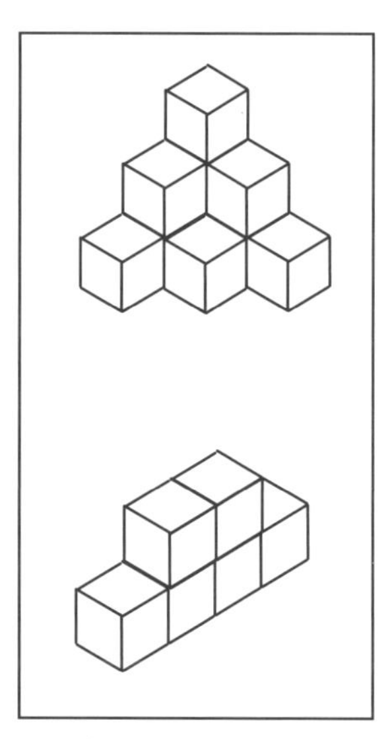

Your student can use unit cubes of any size, such as centimeter cubes, that do not link to each other, or multilink cubes. However, in all of these exercises, assume that the figures drawn on the isometric dot paper can be constructed from blocks that do not link. That is, when a block is not on the lowest level, one or more blocks have to be under it so that no block is suspended in the air. For example, in the pyramid shape to the right, there blocks that are hidden behind other blocks. This figure needs to have 10 blocks in order to construct it, 4 of which are hidden in the drawing but need to be there to support other blocks.

Also assume that there are hidden blocks **only** when one is necessary to support another block. For example, in the 6-block figure at the right, while there could be a block hiding behind the second to the last block in the bottom row, such a block would not be a supporting block, so we know the figure has just the 6 blocks that we see.

Building 3-dimensional models from 2-dimensional pictures can be challenging for some students at this age. Give your student sufficient practice to allow him to interpret pictures of solids such as those on p. 156 of the text well enough to be able to find the volume. If this proves to be too challenging, you may want to save this to do next year prior to the chapter in *Primary Mathematics* 4B on volume. If your student finds the lessons in this chapter easy, you can combine them.

Material

- Unit cubes or multilink cubes
- Appendix pp. a17-a21
- Optional: clay, string

(1) Build solids made of cubes from models

Discussion

Concept p. 151

Show your student an actual cube and have her look at it from different angles. Tell her that the cube takes up space. When we draw a cube on paper, we try to make it look like a real cube, even though the paper is flat. The cubes drawn on both sides of the page are drawn to look like real cubes.

Tasks 1-3, pp. 152

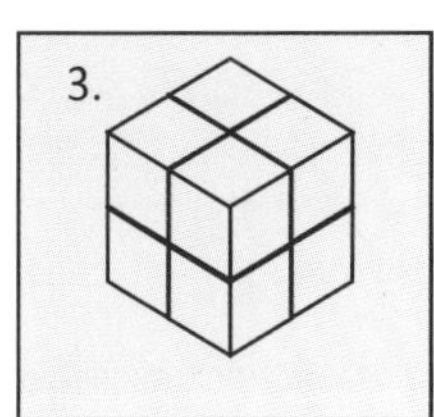

2(a): Make sure your student realizes that the solid is made up of 4 cubes, even though only three are showing. There is a hidden cube in the back. Without it, the figure is not possible; the cubes are not glued to each other along the edges.

2(b): Point out that both solids take up an equal amount of space as 4 unit cubes.

Activity

Students are not required to draw solid figures at this level. If your student likes to draw, you can have him use a copy of the isometric dot paper in the appendix on p. a21 and draw the solid he builds in task 2(a), and perhaps other solids.

Practice

Have your student build the solids on p. a17 in the appendix. Before building the figures, ask how many cubes are needed for each.

p. a17

Number of cubes for each solid:

A: 6 C: 9

B: 6 D: 13

E: 10

F: 10 G: 14

Workbook

Exercise 5, pp. 173-174 (Answers p. 162)

Enrichment

Your student may be interested in knowing some ways to draw cubes or rectangular prisms without the isometric dot paper. A couple of methods are as follows.

To draw a cube with a face front-most, start by drawing a square. Draw parallel lines of equal length backwards and to the side from three of the corners. Connect the ends of these lines with lines parallel to the face.

To draw a cube with an edge front-most, draw two dots a ways apart, and two dots between and just above and below the first two. Connect the dots into a diamond shape. Draw parallel lines of the same length straight down from the two ends and from the near corner. Connect the ends of the lines.

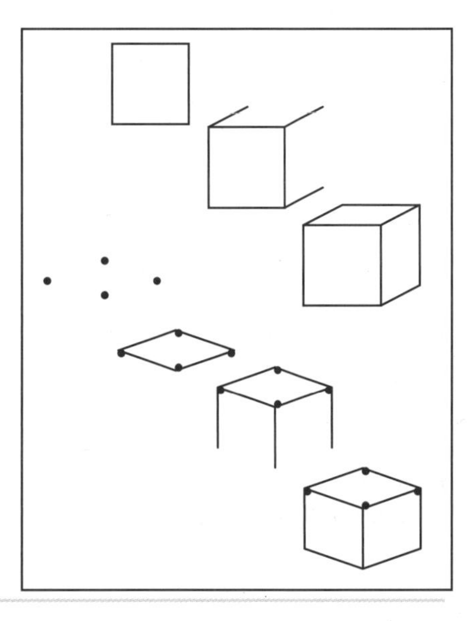

(2) Build solids from models

Discussion

Task 4, p. 153

> 4. 16

Ask your student for the type of solid shown here. It is a rectangular prism. This one can be built out of cubes. We can measure how much space this solid takes up by finding out how many cubes are used to build it. You can have your student build the solid from unit cubes.

Tasks 5-6, pp. 153-154

> 5. A: 5
> B: 6
> C: 9
> D: 9
>
> 6: P: 9
> Q: 8

Note that solids D, P, and Q each have a "hidden" cubes.

Have your student write down her answers. Then ask her which of the 6 solids in these two tasks take up the same amount of space. (C, D, and P). Which one takes up the least amount of space? (A)

Practice

Have your student build the solids on p. a18 in the appendix. Before building the figures, see if he can determine how many cubes are needed for each. Ask which figures take up the same amount of space (A and B) and which is the smallest (F).

> p. a18
> Number of cubes for each solid:
> A: 24 B: 24
> C: 14 D: 10
> E: 21 F: 6

Workbook

Exercise 6, pp. 175-176 (Answers p. 162)

Enrichment

Your student may like to make the solid on p. a20 of the appendix, using multilink cubes. He can try to determine the number of cubes needed in advance.

(3) Add or remove cubes from solid shapes

Discussion

Tasks 7-8, p. 154

7: Have your student build the first solid, and then remove cubes to form the second solid. Ask which solid takes up the most space.

8: Have your student build the first solid and then add cubes to form the second solid. Ask which solid takes up the most space.

7. A: 12 B: 10
2 cubes are removed.

8. C: 4 D: 6
2 cubes are added.

Practice

Have your student build the solids on the left side of p. a19 in the appendix, and either add or remove cubes to make the solids on the right side. Get your student to write down the total number of cubes needed for each solid, and how many were added or removed.

p. a19

A: $24 \xrightarrow{-6} 18$

B: $12 \xrightarrow{+1} 13$

C: $11 \xrightarrow{+10} 21$

Workbook

Exercise 7, pp. 177-178 (Answers p. 162)

Reinforcement

Extra Practice, Unit 13, Exercise 3, pp. 233-236

(Note: In the first printing of this *Extra Practice* book, cm^3 is used for the unit to represent cubic centimeters for problems 3 and 4. Students will not learn this nomenclature until *Primary Mathematics* 4B. You can cross out the cm^3 and simply have your student determine the total number of cubes in the solids.)

(4) Find the volume of solids in cubic units

Activity

Show your student a unit cube. Tell him that the amount of space it takes up is called the **volume** of the cube. We can measure volume in **unit cubes**. If this is the unit we are using, then we can say that the volume of the cube is 1 unit cube.

Give your student 12 unit cubes. Ask him to build a solid out of them. Then ask him for the volume of the solid. It is 12 unit cubes. Ask him to rearrange the 12 cubes into another shape, or make a different solid out of 12 other unit cubes. Ask him if the two shapes have the same volume. They do, since they are both make up of the same number of cubes.

You may want to tell your student that we can use unit cubes that are 1 centimeter on each side or 1 inch on each side. Then we measure volume in cubic centimeters or cubic inches.

Remind your student that some of the areas we measured were not exactly a whole square unit, in which case we used fractions of a square. We can do the same thing with volume. A solid might have a volume of $12\frac{1}{2}$ cubic units. We can also measure irregular shapes, like a sphere. If we say that the volume of a ball is 12 cubic units, then we mean that it takes up the same amount of space as 12 unit cubes. If you have some clay, you can use some thread to cut it into equal-sized cubes. Tell your student that the volume of one of the cubes is a cubic unit. Then roll it into a ball. Its volume is still a cubic unit. Take several of the cubes and form a solid from them without changing their shape and ask your student for the volume. Then form a different shape from the same clay. Tell your student the volume is still the same number of cubic units, even if the shape is no longer box-like.

Discussion

Tasks 9-11, pp. 155-156

Allow your student to build the solids, if needed.

9. **8** cubic units
10. **6** cubic units
11. A: 5 cubic units B: 9 cubic units
C: 18 cubic units D: 12 cubic units
Solid **C** has the greatest volume.

Workbook

Exercise 8, pp. 179 (Answers p. 162)

Test

Tests, Unit 13, 3A and 3B, pp. 293-302

Review 13

Review

Review 13, pp. 157-161

Workbook

Review 13 pp. 180-184 (Answers p. 163)

(Note: Problem 8 is somewhat challenging since the student has not learned a formula for area. But your student should be able to do this by drawing lines to indicate the number of squares there should be. You can draw the lines yourself prior to having him do the review. Or, your student can construct the shapes with unit squares. Let your student determine an approach before offering a solution.)

Test

Tests, Units 1-13, Cumulative A and B, pp. 303-318

1. (a) $\frac{2}{6}, \frac{3}{6}, \frac{5}{6}$ (b) $\frac{4}{9}, \frac{2}{3}, \frac{7}{9}$ (c) $\frac{3}{8}, \frac{1}{2}, \frac{3}{4}$

2. (a) 30 (5 x 6 = 30)
   (b) 10 (40 ÷ 4 = 10)

3.

Figure	A	B	C
Area in square units	6	7	7
Perimeter in units	14	14	16

(a) **B** and **C** have the same area.
(b) **A** and **B** have the same perimeter

4. A: 9 blue, 9 yellow, 18 total
   B: 4 blue, 12 yellow, 16 total
   (a) **18** square units
   (b) **16** square units
   (c) Shape **A** is larger.
   (d) 9 out of 18 = $\frac{9}{18} = \frac{1}{2}$
   $\frac{1}{2}$ of the total square units are blue in A.
   (e) 4 out of 12 = $\frac{4}{16} = \frac{1}{4}$
   $\frac{1}{4}$ of the total square untis are blue in B.

5. **4** cubes were removed.

6. (a) **6** cubes were needed.
   (b) **2** cubes were added.

7. (a) He saved $**8** in January.
   (b) He saved the most money in **March**.
   (c) $8 + $14 + $17 + $14 + $11 = $64
   He saved a total of $**64**.

8.

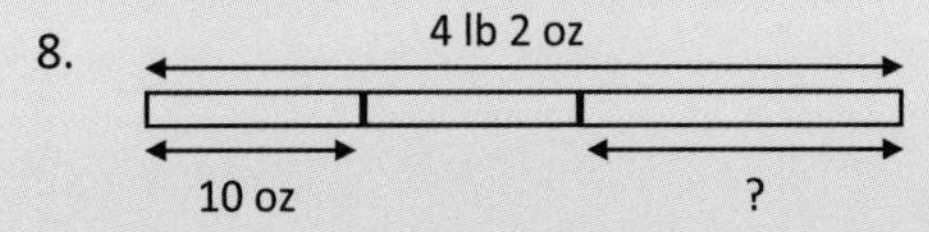

2 bags of sugar weight 20 oz or 1 lb 4 oz.
4 lb 2 oz – 1 lb 4 oz = 2 lb 14 oz
The bag of flour weighs **2 lb 14 oz**.

(Continued next page)

(Textbook Review 13 answers cont.)

9. Width: 5 ft 6 in. ÷ 2
   Method 1: 5 ft 6 in. = 66 in.
   66 in. ÷ 2 = 33 in. = 2 ft 9 in.
   Method 2: Half of 5 ft is 2 ft 6 in.
   Half of 6 in. is 3 in.
   2 ft 6 in. + 3 in. = 2 ft 9 in.
   The table is **2 ft 9 in.** wide.

10. 7 gal – 3 gal 1 qt = 3 gal 3 qt
    3 gal 3 qt of paint was left.

11. 3 pt = 6 c; 1 qt = 4 c, half-gal = 8 c
    Container **D** has the most water.

12. (a) Perimeter: **38 in. or 3 ft 2 in.**
    (b) The figure has **0** right angles.
    (c) The figure has **2** pairs of parallel sides.
    (d) **No**, the figure is not a pentagon.

13. Group $72 by $6.
    $72 ÷ $6 = 12
    She bought **12 lb** of salmon.

14. 2 lb 7 oz = 32 oz + 7 oz = 39 oz
    39 oz ÷ 3 = 13 oz
    Each portion of beans weighed **13 oz**.

15. 25 lb 3 oz – 1 lb 4 oz = 23 lb 15 oz
    The oranges weigh **23 lb 15 oz**.

16. 19 in. = 1 ft 7 in.

17. 29 in. = 2 ft 5 in.; 38 in. = 3 ft 2 in.
    38 in., 3 ft 1 in., 2 ft 9 in., 29 in.
    Or: **D, A, C, B**

18. 18 gal x 4 qt/gal = 72 qt
    The tank can hold **72 qt**.

19. 5 ft 4 in.
    ?
    2 ft 8 in.
    5 ft 4 in. – 2 ft 8 in. = 2 ft 8 in.
    5 ft 4 in. + 2 ft 8 in. = 8 ft 0 in.
    The total length is **8 ft**.

20.
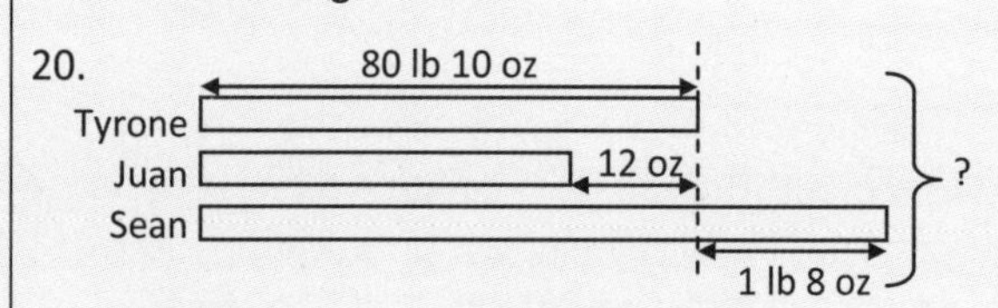

Methods can vary. Possible methods:
Method 1:
80 lb 10 oz – 12 oz = 79 lb 14 oz
80 lb 10 oz + 1 lb 8 oz = 82 lb 2 oz
80 lb 10 oz + 79 lb 14 oz + 82 lb 2 oz
= 241 lb 10 oz + 16 oz = 242 lb 10 oz
Method 2:
80 lb 10 oz x 3 = 240 lb 30 oz
1 lb 8 oz – 12 oz = 12 oz
240 lb 30 oz + 12 oz
= 240 lb + 32 oz + 10 oz = 242 lb 10 oz
The total weight is **242 lb 10 oz**.

21. (a) likely
    (b) certain
    (c) unlikely
    (d) impossible

Workbook

Exercise 1, pp, 159-162

1. (a) (b) (c) (d) (e)

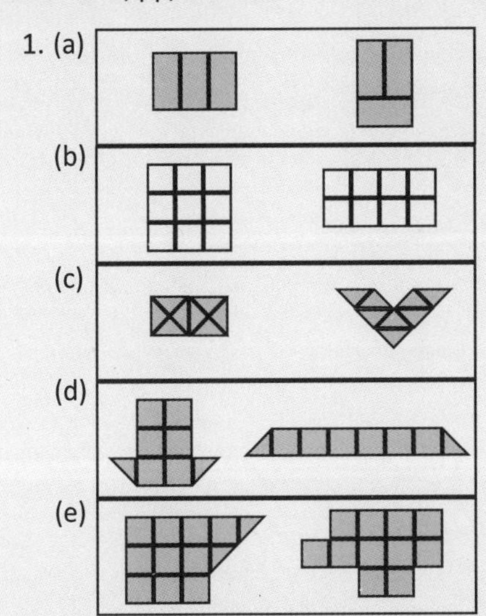

2. (a) (b) (c) (d) (e)

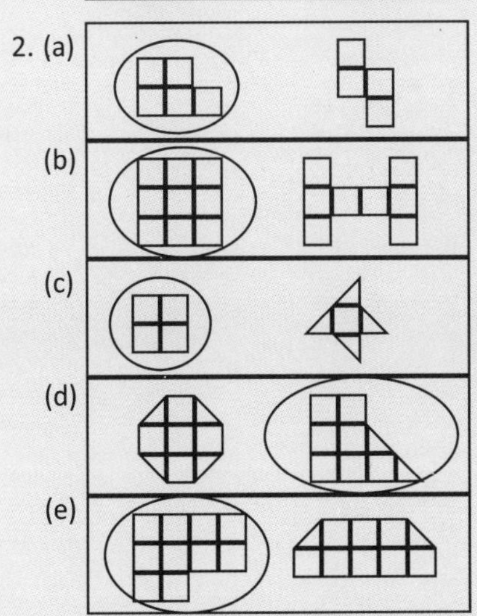

3. Answers will vary; check answers.

4.

Figure	Area
A	9 square units
B	10 square units
C	9 square units
D	6 square units
E	9 square units
F	10 square units

Exercise 2, pp. 163-166

1.

2. (a) 8 (b) 5
(c) 7 (d) 9
(e) 9 (f) 16

3. Answers will vary; check answers.

4. (a)

Figure	Area
A	6 square units
B	5 square units
C	6 square units
D	7 square units

(b) Shape **D** has the greatest area.
(c) Shape **B** has the smallest area.
(d) Shape **A** and Shape **C** have the same area.

Exercise 3, pp. 167-169

1.

Figure	Area
A	11 square centimeters
B	11 square centimeters
C	10 square centimeters
D	13 square centimeters

Figure **A** and Figure **B** have the same area.
Figure **D** has the biggest area.
Figure **C** has the smallest area.

2. Answers will vary; check answers.
Areas: 9 square centimeters
5 square centimeters
7 square centimeters
8 square centimeters

3. (a) 8 square inches (b) 12 square inches
(c) 5 square inches (d) 6 square inches

Workbook

Exercise 4, pp. 170-172

1. (a) about 14 cm (b) 16 cm
   (c) 12 cm (d) 16 cm
   (e) about 10 cm (f) 12 cm

2. (a)

Figure	A	B	C	D	E	F
Area in square centimeters	10	13	10	9	8	13
Perimeter in centimeters	14	16	16	12	18	16

   (b) A and C
   (c) B and C or C and F
   (d) B and F

3. (a) 28 cm
   (b) 36 cm
   (c) 34 in.
   (d) 37 in.

Exercise 5, pp. 173-174

1. (a) 3
   (b) 5
   (c) 7

2.

Solid	Number of unit cubes
A	3
B	4
C	2
D	6
E	5
F	8

Exercise 6, pp. 175-176

2.

Solid	A	B	C	D	E
Number of unit cubes	16	27	6	9	7

Exercise 7, pp. 177-178

1. (a) 6 ⟶ 5 [−1]
   (b) 7 ⟶ 5 [−2]
   (a) 9 ⟶ 7 [−2]

2. (a) 4 ⟶ 5 [+1]
   (b) 4 ⟶ 6 [+2]
   (a) 5 ⟶ 7 [+2]

Exercise 8, p. 179

1. A: 12 B: 6
   C: 16 D: 15
   E: 10 F: 15
   C has the greatest volume.
   B has the smallest volume.

Workbook

Review 13, pp. 180-184

1. (a) 5000 (b) 50

2. (a) $\frac{\mathbf{6}}{9}$ (b) $\frac{3}{\mathbf{5}}$ (c) $\frac{\mathbf{5}}{6}$

3. (a) $\frac{4}{5}$ (b) $\frac{3}{4}$ (c) $\frac{1}{2}$

4. (a) $\frac{1}{8}, \frac{4}{8}, \frac{7}{8}$
   (b) $\frac{2}{15}, \frac{5}{15}, \frac{4}{5}$
   (c) $\frac{2}{8}, \frac{2}{6}, \frac{1}{2}$ (Either simplify the first two fractions, and then all three have the same numerator, or notice that the first two both have to be less than a half, and already have the same numerator.)

5. (a) 40 (8 x 5)
   (b) 30 (90 ÷ 3)

6. (a) Area = 11 sq. cm
   Perimeter = 16 cm
   (b) Area = 7 sq. cm
   Perimeter = 14 cm

7. (a) 35 cm (b) 36 in.

8. (a) Area of rectangle is 15 square centimeters.
   (b) Perimeter of rectangle: 16 cm
   So perimeter of square: 16 cm
   Side of square: 16 cm ÷ 4 = 4 cm
   Draw squares to see that there are 16.
   Area of the square is 16 square centimeters

9. There are 4 250's in 1000, and 2 in 500. So
   6 x 250 ml = 1 ℓ 500 ml
   Or: 6 x 250 ml = 1500 ml = **1 ℓ 500 ml**

10. 2 km 750 m – 1 km 900 m = **850 m**

11. P: 11 cubes; Q: 7 cubes. So difference is 4 cubes.
    Or: To go from P to Q, remove 2 cubes from left side, Move the top one at the right over to the other side and remove the 2 at the end. 4 cubes were removed.
    The difference in volume is **4 cubic units**.

12. 2 right angles

13. $4.50 x 8 = 450¢ x 8 = 3600¢ = $36.00
    The book costs **$36**.

14. He left his house at **6:50 a.m.**

15. 1 box: 8 + 6 = 14
    5 boxes = 14 x 5 = 70
    There are **70** cookies in 5 boxes.

16. (a) 3 lb 3 oz
    (b) 16 pt 1 c
    (c) 4 ft 6 in.

17. (a) She spent $\frac{3}{4}$ of a dollar on candy.
    (b) 3 quarters = 75¢
    Total spent: $3.60 + $0.75 = $4.35
    Total left: $6.00 – $4.35 = $1.65
    She had **$1.65** left.

Mental Math 1	Mental Math 2	Mental Math 3	Mental Math 4	Mental Math 5	Mental Math 6
100 – 50 = **50**	1000 – 750 = **250**	1000 – 7 = **993**	2000 – 480 = **1520**	12 x 1 = **12**	3 ft = **1** yd
200 – 50 = **150**	1000 – 400 = **600**	1000 – 348 = **652**	5000 – 12 = **4988**	12 x 5 = **60**	36 ft = **12** yd
100 – 35 = **65**	1000 – 350 = **650**	1000 – 830 = **170**	3000 – 35 = **2965**	12 x 2 = **24**	6 ft = **2** yd
600 – 35 = **565**	1000 – 420 = **580**	1000 – 769 = **231**	6000 – 740 = **5260**	12 x 8 = **96**	120 in. = **10** ft
100 – 48 = **52**	1000 – 560 = **440**	1000 – 910 = **90**	7000 – 12 = **6988**	12 x 4 = **48**	3 yd = **9** ft
500 – 48 = **452**	1000 – 107= **893**	1000 – 375 = **625**	9000 – 581 = **8419**	12 x 7 = **84**	6 ft = **72** in.
100 – 32 = **68**	1000 – 238= **762**	1000 – 603 = **397**	4000 – 263 = **3737**	12 x 6 = **72**	24 yd = **72** ft
700 – 86 = **614**	1000 – 423= **577**	1000 – 92 = **908**	8000 – 435 = **7565**	12 x 8 = **96**	36 in. = **3** ft
900 – 43 = **857**	1000 – 126 = **874**	1000 – 707 = **293**	2000 – 947 = **1053**	12 x 3 = **36**	9 ft = **108** in.
800 – 92 = **708**	1000 – 514 = **486**	1000 – 955 = **45**	4000 – 678 = **3322**	12 x 10 = **120**	12 yd = **36** ft
200 – 56 = **144**	1000 – 379 = **621**	1000 – 28 = **972**	6000 – 826 = **5174**	48 ÷ 12 = **4**	9 yd = **27** ft
700 – 61 = **639**	1000 – 765 = **235**	1000 – 333 = **667**	3000 – 199 = **2801**	84 ÷ 12 = **7**	120 ft = **40** yd
400 – 24 = **376**	1000 – 88 = **912**	1000 – 19 = **981**	9000 – 230 = **8770**	24 ÷ 12 = **2**	3 ft = **36** in.
300 – 87 = **213**	1000 – 632 = **368**	1000 – 743 = **257**	7000 – 81 = **6919**	72 ÷ 12 = **6**	6 yd = **18** ft
600 – 79 = **521**	1000 – 904 = **96**	1000 – 450 = **550**	1000 – 694 = **306**	120 ÷ 12 = **10**	12 in. = **1** ft
500 – 32 = **468**	1000 – 190 = **810**	1000 – 364 = **636**	5000 – 352 = **4648**	12 ÷ 12 = **1**	3 yd = **108** in.
900 – 68 = **832**	1000 – 41 = **959**	1000 – 802 = **198**	9000 – 43 = **8957**	60 ÷ 12 = **5**	12 ft = **4** yd
400 – 23 = **377**	1000 – 153 = **847**	1000 – 970 = **30**	3000 – 540 = **2460**	36 ÷ 12 = **3**	24 in. = **2** ft
800 – 85 = **715**	1000 – 16 = **984**	1000 – 86 = **914**	6000 – 485 = **5515**	108 ÷ 12 = **9**	36 in. = **1** yd
300 – 14 = **286**	1000 – 931 = **69**	1000 – 197 = **803**	5000 – 777 = **4223**	96 ÷ 12 = **8**	120 yd = **360** ft

Mental Math 7	Mental Math 8	Mental Math 9	Mental Math 10
5 lb – 5 oz = **4** lb **11** oz	4 qt = **8** pt	$4.50 + 90¢ = $**5.40**	$6.00 – 40¢ = $**5.60**
3 yd – 2 ft = **2** yd **1** ft	9 gal = **36** qt	$7.85 + 80¢ = $**8.65**	$9.20 – 30¢ = $**8.90**
4 ft – 4 in. = **3** ft **8** in.	2 pt = **4** c	$9.35 + 20¢ = $**9.55**	$2.05 – 10¢ = $**1.95**
2 m – 8 cm = **1** m **92** cm	6 km = **6000** m	$7.90 + 60¢ = $**8.50**	$8.55 – 50¢ = $**8.05**
9 km – 7 m = **8** km **993** m	1 gal = **8** pt	$8.25 + 40¢ = $**8.65**	$3.25 – 15¢ = $**3.10**
8 kg – 3 g = **7** kg **997** g	8 yd = **24** ft	$4.30 + 95¢ = $**5.25**	$8.30 – 60¢ = $**7.70**
6 lb – 6 oz = **5** lb **10** oz	2 qt = **8** c	$8.65 + 65¢ = $**9.30**	$3.65 – 80¢ = $**2.85**
9 oz + 8 oz = **1** lb **1** oz	7 m = **700** cm	$6.50 + 75¢ = $**7.25**	$7.40 – 75¢ = $**6.65**
900 m + 800 m = **1** km **700** m	3 km = **3000** m	$8.05 + 40¢ = $**8.45**	$8.70 – 25¢ = $**8.45**
99 cm + 8 cm = **1** m **7** cm	9 yd = **27** ft	$9.55 + 85¢ = $**10.40**	$4.35 – 60¢ = $**3.75**
9 in. + 8 in. = **1** ft **5** in.	5 qt = **20** c	$5.20 + 75¢ = $**5.95**	$89.05 – 45¢ = $**88.60**
999 g + 99 g = **1** kg **98** g	6 m = **600** cm	$4.85 + 35¢ = $**5.20**	$42.70 – 15¢ = $**42.55**
5 yd 2 ft + 2 ft = **6** y **1** ft	3 kg = **3000** g	$13.40 + 70¢ = $**14.10**	$37.25 – 50¢ = $**36.75**
3 km 70 m + 60 m = **3** km **130** m	2 ft = **24** in.	$36.75 + 40¢ = $**37.15**	$59.55 – 75¢ = $**58.80**
3 m 70 cm + 60 cm = **4** m **30** cm	10 lb = **160** oz	$49.50 + 55¢ = $**50.05**	$24.80 – 35¢ = $**24.45**
3 lb 7 oz + 11 oz = **4** lb **2** oz	5 gal = **40** pt	$34.55 + 80¢ = $**35.35**	$14.55 – 90¢ = $**13.65**
3 ft 7 in. + 11 in. = **4** ft **6** in.	6 yd = **18** ft	$69.25 + 60¢ = $**69.85**	$13.40 – 95¢ = $**12.45**
3 kg 7 g + 11 g = **3** kg **18** g	8 kg = **8000** g	$63.90 + 95¢ = $**64.85**	$75.75 – 99¢ = $**74.76**
8 lb 7 oz + 6 oz = **8** lb **13** oz	2 gal = **32** c	$15.35 + 98¢ = $**16.33**	$52.05 – 98¢ = $**51.07**
3 ft + 8 ft = **3** y **2** ft	7 ft = **84** in.	$45.85 + 97¢ = $**46.82**	$29.90 – 96¢ = $**28.94**

Mental Math 11	Mental Math 12	Mental Math 13	Mental Math 14
$47.80 + 15¢ = **$47.95**	1 h – 30 min = **30** min	8 min = **480** seconds	Fill in the circle with >, <, or =
$13.25 + 95¢ = **$14.20**	1 h – 15 min = **45** min	5 years = **60** months	9 weeks **>** 56 days
$85.05 – 99¢ = **$84.06**	1 h – 45 min = **15** min	7 min = **420** seconds	2 months **<** 10 weeks
$54.95 + 25¢ = **$55.20**	1 h – 55 min = **5** min	36 months = **3** years	60 months **<** 6 years
$96.35 – 65¢ = **$95.70**	1 h – 35 min = **25** min	140 days = **20** weeks	1 h – 25 min **>** 25 min
$60.20 – 40¢ = **$59.80**	1 h – 50 min = **10** min	10 years = **120** months	52 s + 44 s **<** 2 min
$45.55 + 35¢ = **$45.90**	1 h – 25 min = **35** min	6 hours = **360** minutes	140 months **<** 10 years 40 months
$15.95 + 20¢ = **$16.15**	1 h – 40 min = **20** min	49 days = **7** weeks	3 days **>** 70 h
$43.20 – 91¢ = **$42.29**	1 h – 10 min = **50** min	96 months = **8** years	4 h 65 s **>** 241 min
$64.45 – 96¢ = **$63.49**	1 h – 20 min = **40** min	2 years = **24** months	3 h – 45 min **=** 2 h 45 min – 30 min
$28.35 + 45¢ = **$28.80**	1 h – 5 min = **55** min	4 weeks = **28** days	6 months – 6 weeks **>** 4 months
$34.50 + 75¢ = **$35.25**	1 h – 42 min = **18** min	3 hours = **180** minutes	372 days **>** 53 weeks
$83.20 – 97¢ = **$82.23**	1 h – 8 min = **52** min	9 min = **540** seconds	3600 s **>** 36 min
$36.45 + 75¢ = **$37.20**	1 h – 27 min = **33** min	48 months = **4** years	6 weeks 19 days **>** 8 weeks
$48.90 + 97¢ = **$49.87**	1 h – 58 min = **2** min	35 days = **5** weeks	2 h – 120 s **>** 1 h + 40 min
$25.95 + 40¢ = **$26.35**	1 h – 16 min = **44** min	20 hours = **1200** minutes	Half a year **=** 26 weeks
$63.25 – 80¢ = **$62.45**	1 h – 33 min = **27** min	12 years = **144** months	30 days **<** 2 months – 3 weeks
$19.00 – 5¢ = **$18.95**	1 h – 26 min = **34** min	84 months = **7** years	
$43.09 – 99¢ = **$42.10**	1 h – 11 min = **49** min	14 days = **2** weeks	
$50.60 – 65¢ = **$49.95**	1 h – 52 min = **8** min	5 min = **300** seconds	

Blank page

Mental Math 1	Mental Math 2	Mental Math 3
100 – 50 = ______	1000 – 750 = ______	1000 – 7 = ______
200 – 50 = ______	1000 – 400 = ______	1000 – 348 = ______
100 – 35 = ______	1000 – 350 = ______	1000 – 830 = ______
600 – 35 = ______	1000 – 420 = ______	1000 – 769 = ______
100 – 48 = ______	1000 – 560 = ______	1000 – 910 = ______
500 – 48 = ______	1000 – 107= ______	1000 – 375 = ______
100 – 32 = ______	1000 – 238= ______	1000 – 603 = ______
700 – 86 = ______	1000 – 423= ______	1000 – 92 = ______
900 – 43 = ______	1000 – 126 = ______	1000 – 707 = ______
800 – 92 = ______	1000 – 514 = ______	1000 – 955 = ______
200 – 56 = ______	1000 – 379 = ______	1000 – 28 = ______
700 – 61 = ______	1000 – 765 = ______	1000 – 333 = ______
400 – 24 = ______	1000 – 88 = ______	1000 – 19 = ______
300 – 87 = ______	1000 – 632 = ______	1000 – 743 = ______
600 – 79 = ______	1000 – 904 = ______	1000 – 450 = ______
500 – 32 = ______	1000 – 190 = ______	1000 – 364 = ______
900 – 68 = ______	1000 – 41 = ______	1000 – 802 = ______
400 – 23 = ______	1000 – 153 = ______	1000 – 970 = ______
800 – 85 = ______	1000 – 16 = ______	1000 – 86 = ______
300 – 14 = ______	1000 – 931 = ______	1000 – 197 = ______

Mental Math 4	Mental Math 5	Mental Math 6
2000 – 480 = ______	12 x 1 = ______	3 ft = ______ yd
5000 – 12 = ______	12 x 5 = ______	36 ft = ______ yd
3000 – 35= ______	12 x 2 = ______	6 ft = ______ yd
6000 – 740 = ______	12 x 8 = ______	120 in. = ______ ft
7000 – 12 = ______	12 x 4 = ______	3 yd = ______ ft
9000 – 581 = ______	12 x 7 = ______	6 ft = ______ in.
4000 – 263 = ______	12 x 6 = ______	24 yd = ______ft
8000 – 435 = ______	12 x 8 = ______	36 in. = ______ ft
2000 – 947 = ______	12 x 3 = ______	9 ft = ______ in.
4000 – 678 = ______	12 x 10 = ______	12 yd = ______ft
6000 – 826 = ______	48 ÷ 12 = ______	9 yd = ______ft
3000 – 199 = ______	84 ÷ 12 = ______	120 ft = ______ yd
9000 – 230 = ______	24 ÷ 12 = ______	3 ft = ______ in.
7000 – 81 = ______	72 ÷ 12 = ______	6 yd = ______ft
1000 – 694 = ______	120 ÷ 12 = ______	12 in. = ______ ft
5000 – 352 = ______	12 ÷ 12 = ______	3 yd = ______ in.
9000 – 43 = ______	60 ÷ 12 = ______	12 ft = ______ yd
3000 – 540 = ______	36 ÷ 12 = ______	24 in. = ______ ft
6000 – 485 = ______	108 ÷ 12 = ______	36 in. = ______ yd
5000 – 777 = ______	96 ÷ 12 = ______	120 yd = ______ ft

Mental Math 7	Mental Math 8
5 lb – 5 oz = ______ lb ______ oz	4 qt = ______ pt
3 yd – 2 ft = ______ yd ______ ft	9 gal = ______ qt
4 ft – 4 in. = ______ ft ______ in.	2 pt = ______ c
2 m – 8 cm = ______ m ______ cm	6 km = ______ m
9 km – 7 m = ______ km ______ m	1 gal = ______ pt
8 kg – 3 g = ______ kg ______ g	8 yd = ______ ft
6 lb – 6 oz = ______ lb ______ oz	2 qt = ______ c
9 oz + 8 oz = ______ lb ______ oz	7 m = ______ cm
900 m + 800 m = ______ km ______ m	3 km = ______ m
99 cm + 8 cm = ______ m ______ cm	9 yd = ______ ft
9 in. + 8 in. = ______ ft ______ in.	5 qt = ______ c
999 g + 99 g = ______ kg ______ g	6 m = ______ cm
5 yd 2 ft + 2 ft = ______ y ______ ft	3 kg = ______ g
3 km 70 m + 60 m = ______ km ______ m	2 ft = ______ in.
3 m 70 cm + 60 cm = _____ m _____ cm	10 lb = ______ oz
3 lb 7 oz + 11 oz = ______ lb ______ oz	5 gal = ______ pt
3 ft 7 In. + 11 in. = ______ ft ______ in.	6 yd = ______ ft
3 kg 7 g + 11 g = ______ kg ______ g	8 kg = ______ g
8 lb 7 oz + 6 oz = ______ lb ______ oz	2 gal = ______ c
3 ft + 8 ft = ______ y ______ ft	7 ft = ______ in.

Mental Math 9	Mental Math 10
\$4.50 + 90¢ = \$________	\$6.00 – 40¢ = \$________
\$7.85 + 80¢ = \$________	\$9.20 – 30¢ = \$________
\$9.35 + 20¢ = \$________	\$2.05 – 10¢ = \$________
\$7.90 + 60¢ = \$________	\$8.55 – 50¢ = \$________
\$8.25 + 40¢ = \$________	\$3.25 – 15¢ = \$________
\$4.30 + 95¢ = \$________	\$8.30 – 60¢ = \$________
\$8.65 + 65¢ = \$________	\$3.65 – 80¢ = \$________
\$6.50 + 75¢ = \$________	\$7.40 – 75¢ = \$________
\$8.05 + 40¢ = \$________	\$8.70 – 25¢ = \$________
\$9.55 + 85¢ = \$________	\$4.35 – 60¢ = \$________
\$5.20 + 75¢ = \$________	\$89.05 – 45¢ = \$________
\$4.85 + 35¢ = \$________	\$42.70 – 15¢ = \$________
\$13.40 + 70¢ = \$________	\$37.25 – 50¢ = \$________
\$36.75 + 40¢ = \$________	\$59.55 – 75¢ = \$________
\$49.50 + 55¢ = \$________	\$24.80 – 35¢ = \$________
\$34.55 + 80¢ = \$________	\$14.55 – 90¢ = \$________
\$69.25 + 60¢ = \$________	\$13.40 – 95¢ = \$________
\$63.90 + 95¢ = \$________	\$75.75 – 99¢ = \$________
\$15.35 + 98¢ = \$________	\$52.05 – 98¢ = \$________
\$45.85 + 97¢ = \$________	\$29.90 – 96¢ = \$________

Mental Math 11	Mental Math 12
\$47.80 + 15¢ = \$________	1 h – 30 min = ________ min
\$13.25 + 95¢ = \$________	1 h – 15 min = ________ min
\$85.05 – 99¢ = \$________	1 h – 45 min = ________ min
\$54.95 + 25¢ = \$________	1 h – 55 min = ________ min
\$96.35 – 65¢ = \$________	1 h – 35 min = ________ min
\$60.20 – 40¢ = \$________	1 h – 50 min = ________ min
\$45.55 + 35¢ = \$________	1 h – 25 min = ________ min
\$15.95 + 20¢ = \$________	1 h – 40 min = ________ min
\$43.20 – 91¢ = \$________	1 h – 10 min = ________ min
\$64.45 – 96¢ = \$________	1 h – 20 min = ________ min
\$28.35 + 45¢ = \$________	1 h – 5 min = ________ min
\$34.50 + 75¢ = \$________	1 h – 42 min = ________ min
\$83.20 – 97¢ = \$________	1 h – 8 min = ________ min
\$36.45 + 75¢ = \$________	1 h – 27 min = ________ min
\$48.90 + 97¢ = \$________	1 h – 58 min = ________ min
\$25.95 + 40¢ = \$________	1 h – 16 min = ________ min
\$63.25 – 80¢ = \$________	1 h – 33 min = ________ min
\$19.00 – 5¢ = \$________	1 h – 26 min = ________ min
\$43.09 – 99¢ = \$________	1 h – 11 min = ________ min
\$50.60 – 65¢ = \$________	1 h – 52 min = ________ min

Mental Math 13

8 min = ______ seconds

5 years = ______ months

7 min = ______ seconds

36 months = ______ years

140 days = ______ weeks

10 years = ______ months

6 hours = ______ minutes

49 days = ______ weeks

96 months = ______ years

2 years = ______ months

4 weeks = ______ days

3 hours = ______ minutes

9 min = ______ seconds

48 months = ______years

35 days = ______weeks

20 hours = ______ minutes

12 years = ______ months

84 months = ______years

14 days = ______ weeks

5 min = ______ seconds

Mental Math 14

Fill in the circle with >, <, or =

9 weeks ○ 56 days

2 months ○ 10 weeks

60 months ○ 6 years

1 h – 25 min ○ 25 min

52 s + 44 s ○ 2 min

140 months ○ 10 years 40 months

3 days ○ 70 h

4 h 65 s ○ 241 min

3 h – 45 min ○ 2 h 45 min – 30 min

6 months – 6 weeks ○ 4 months

372 days ○ 53 weeks

3600 s ○ 36 min

6 weeks 19 days ○ 8 weeks

2 h – 120 s ○ 1 h + 40 min

Half a year ○ 26 weeks

30 days ○ 2 months – 3 weeks

1. Put >, <, or = in the circles.

(a) 10 yd 2 ft ◯ 22 ft (b) 12 yd ◯ 144 ft

(c) 15 ft ◯ 3 yd (d) 3 yd ◯ 3 m

(e) 120 ft ◯ 10 yd (f) 4 yd 2 ft ◯ 14 ft

2. Study the treasure map (X marks the treasure) and answer the questions below.

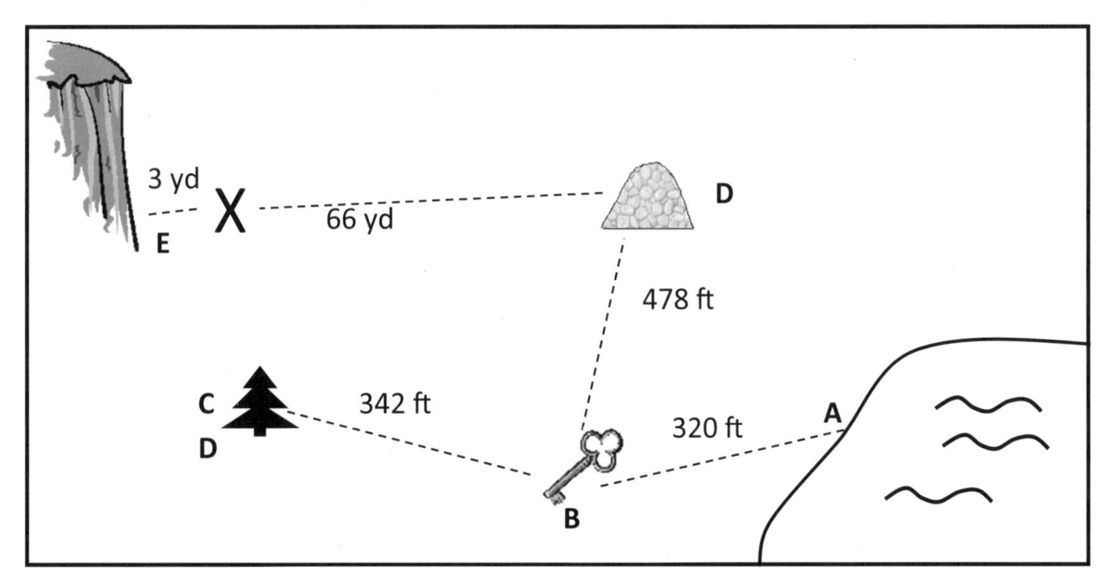

(a) What is the distance from the shore (A) to the pine tree (C) in yards and feet?

(b) What is the distance from the key (B) to the treasure, going by way of the pile of rocks (D) in feet?

(c) How far is the treasure from the foot of the cliff in feet?

(d) A pirate hiked from the shore (A) to the key (B) and then to the rock pile (D) where he was ambushed. Give the distance he hiked in yards and feet.

1. Add in compound units

   (a) 5 yd 2 ft + 2 ft = ______ yd ______ ft

   (b) 6 yd 2 ft + 4 yd 2 ft = ______ yd ______ ft

   (c) 8 yd 1 ft + 12 yd 1 ft = ______ yd ______ ft

   (d) 15 yd 1 ft + 3 yd 2 ft = ______ yd ______ ft

2. Subtract in compound units

   (a) 10 yd – 2 ft = ______ ft

   (b) 8 yd 1 ft – 2 ft = ______ yd ______ ft

   (c) 7 yd 2 ft – 5 yd 1 ft = ______ yd ______ ft

   (d) 12 yd 2 ft – 5 yd = ______ yd ______ ft

   (e) 18 yd 2 ft – 6 yd 2 ft = ______ yd ______ ft

   (f) 14 yd 1 ft – 4 yd 2 ft = ______ yd ______ ft

3. Brett had two sticks. The first one was 4 yd 1 ft long and the second one was 2 yd 2 ft long.

   (a) What is their total length?

   (b) How much longer is the first one that the second one?

$\frac{1}{1}$

$\frac{1}{2}$ $\frac{1}{2}$

$\frac{1}{3}$ $\frac{1}{3}$ $\frac{1}{3}$

$\frac{1}{4}$ $\frac{1}{4}$ $\frac{1}{4}$ $\frac{1}{4}$

$\frac{1}{5}$ $\frac{1}{5}$ $\frac{1}{5}$ $\frac{1}{5}$ $\frac{1}{5}$

$\frac{1}{6}$ $\frac{1}{6}$ $\frac{1}{6}$ $\frac{1}{6}$ $\frac{1}{6}$ $\frac{1}{6}$

$\frac{1}{7}$ $\frac{1}{7}$ $\frac{1}{7}$ $\frac{1}{7}$ $\frac{1}{7}$ $\frac{1}{7}$ $\frac{1}{7}$

$\frac{1}{8}$ $\frac{1}{8}$ $\frac{1}{8}$ $\frac{1}{8}$ $\frac{1}{8}$ $\frac{1}{8}$ $\frac{1}{8}$ $\frac{1}{8}$

$\frac{1}{9}$ $\frac{1}{9}$ $\frac{1}{9}$ $\frac{1}{9}$ $\frac{1}{9}$ $\frac{1}{9}$ $\frac{1}{9}$ $\frac{1}{9}$ $\frac{1}{9}$

$\frac{1}{10}$ $\frac{1}{10}$ $\frac{1}{10}$ $\frac{1}{10}$ $\frac{1}{10}$ $\frac{1}{10}$ $\frac{1}{10}$ $\frac{1}{10}$ $\frac{1}{10}$ $\frac{1}{10}$

$\frac{1}{11}$ $\frac{1}{11}$ $\frac{1}{11}$ $\frac{1}{11}$ $\frac{1}{11}$ $\frac{1}{11}$ $\frac{1}{11}$ $\frac{1}{11}$ $\frac{1}{11}$ $\frac{1}{11}$ $\frac{1}{11}$

$\frac{1}{12}$ $\frac{1}{12}$ $\frac{1}{12}$ $\frac{1}{12}$ $\frac{1}{12}$ $\frac{1}{12}$ $\frac{1}{12}$ $\frac{1}{12}$ $\frac{1}{12}$ $\frac{1}{12}$ $\frac{1}{12}$ $\frac{1}{12}$

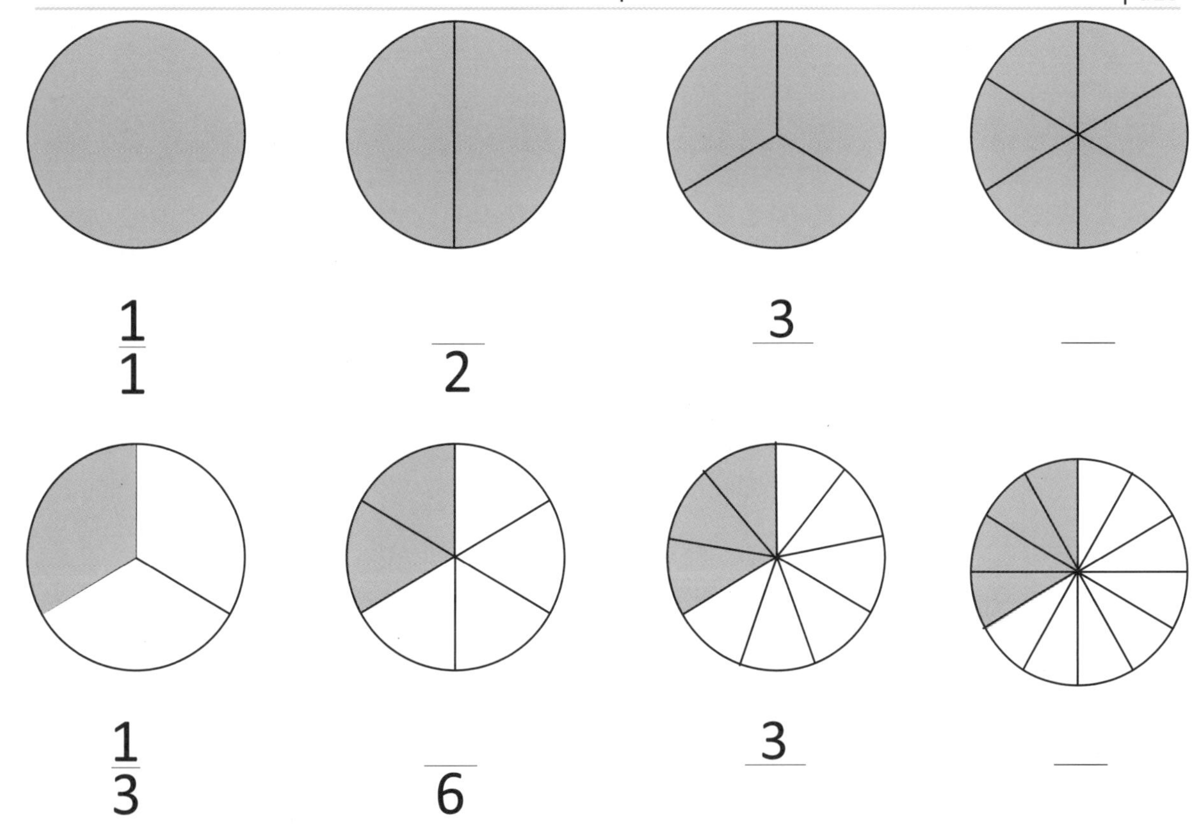
1/1
2
3
1/3
6
3

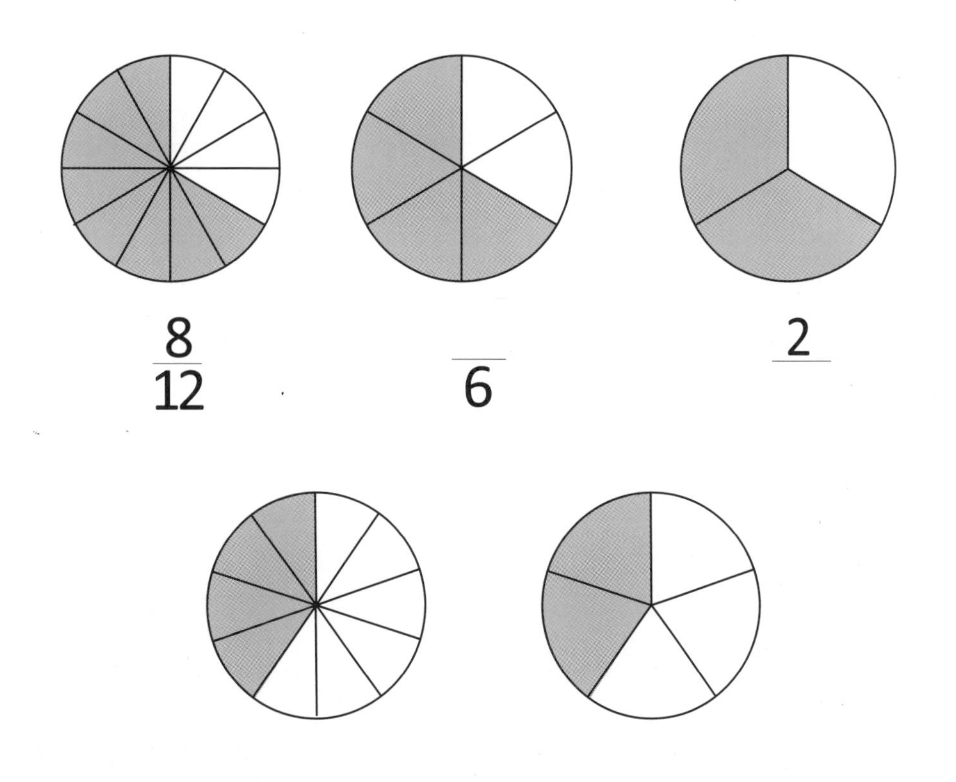
8/12
6
2

Textbook p. 147

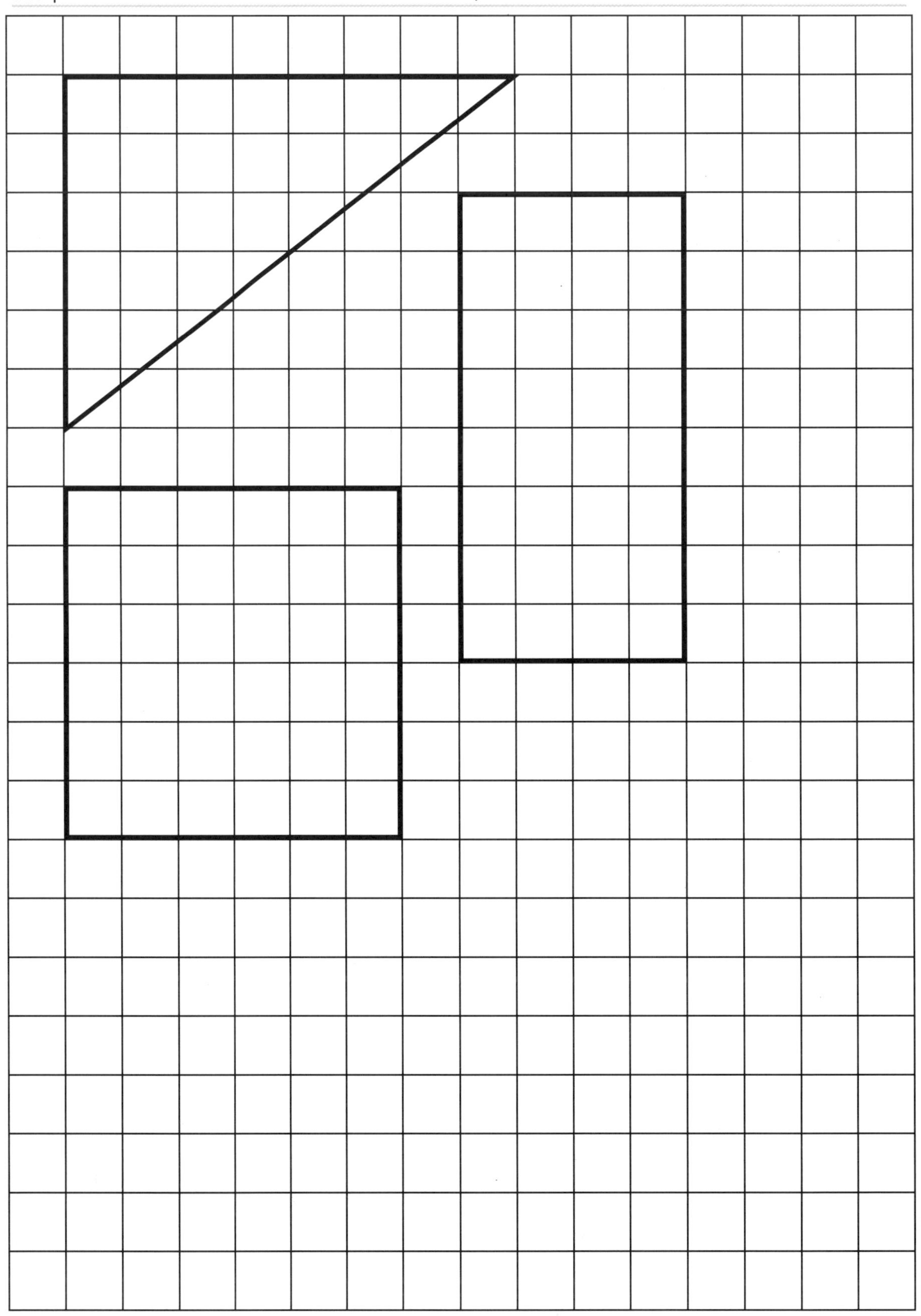

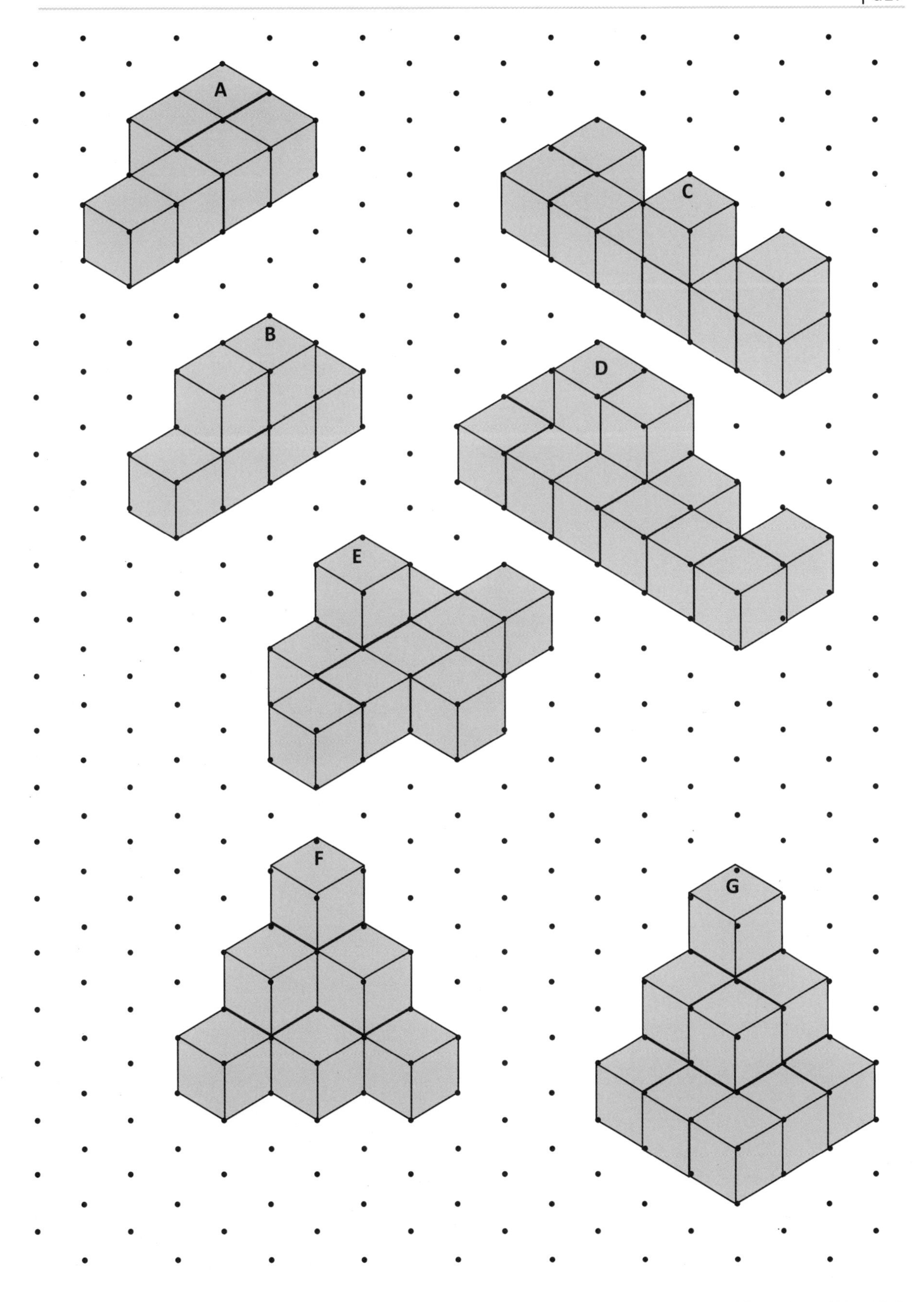
A
B
C
D
E
F
G

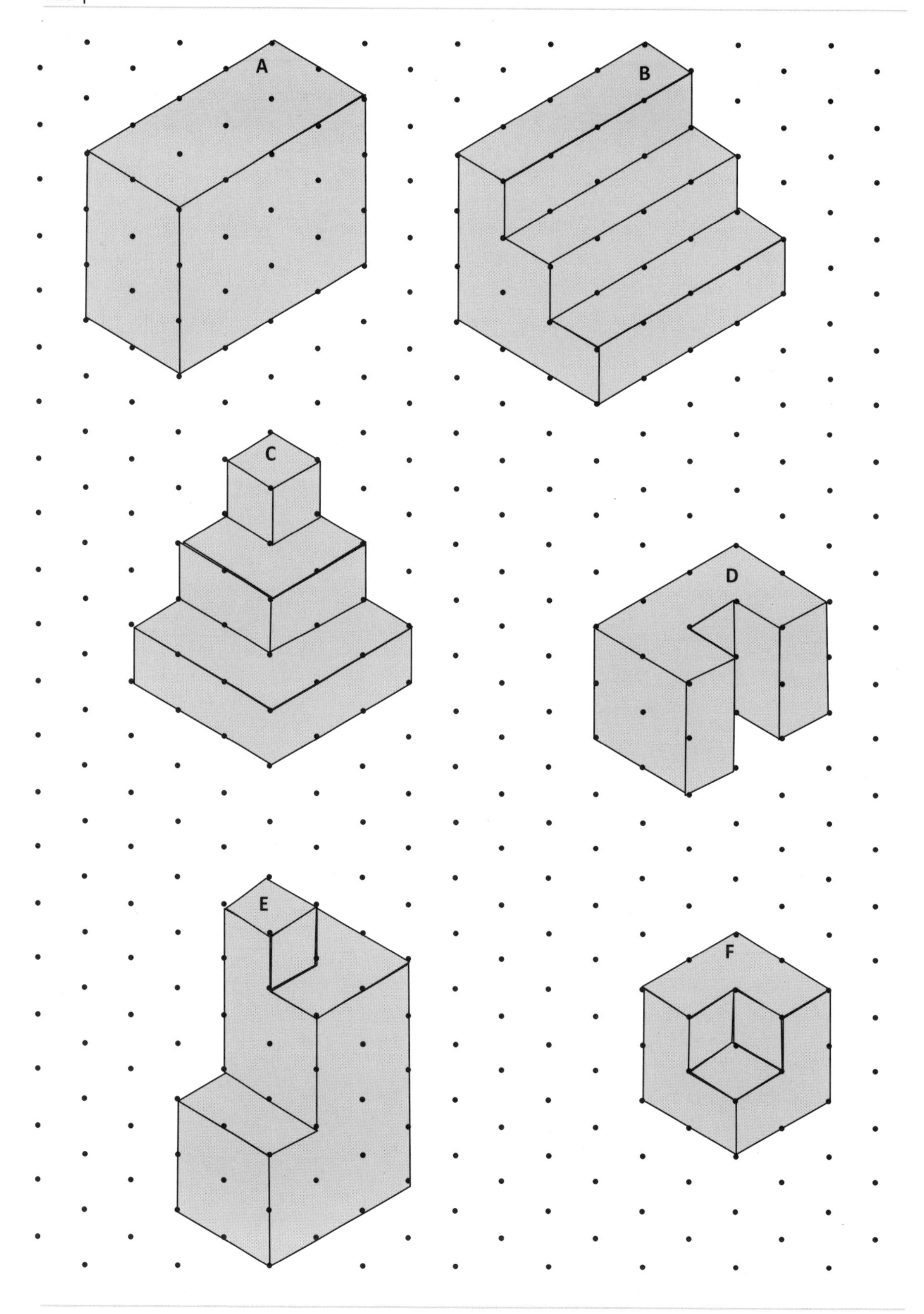
A
B
C
D
E
F

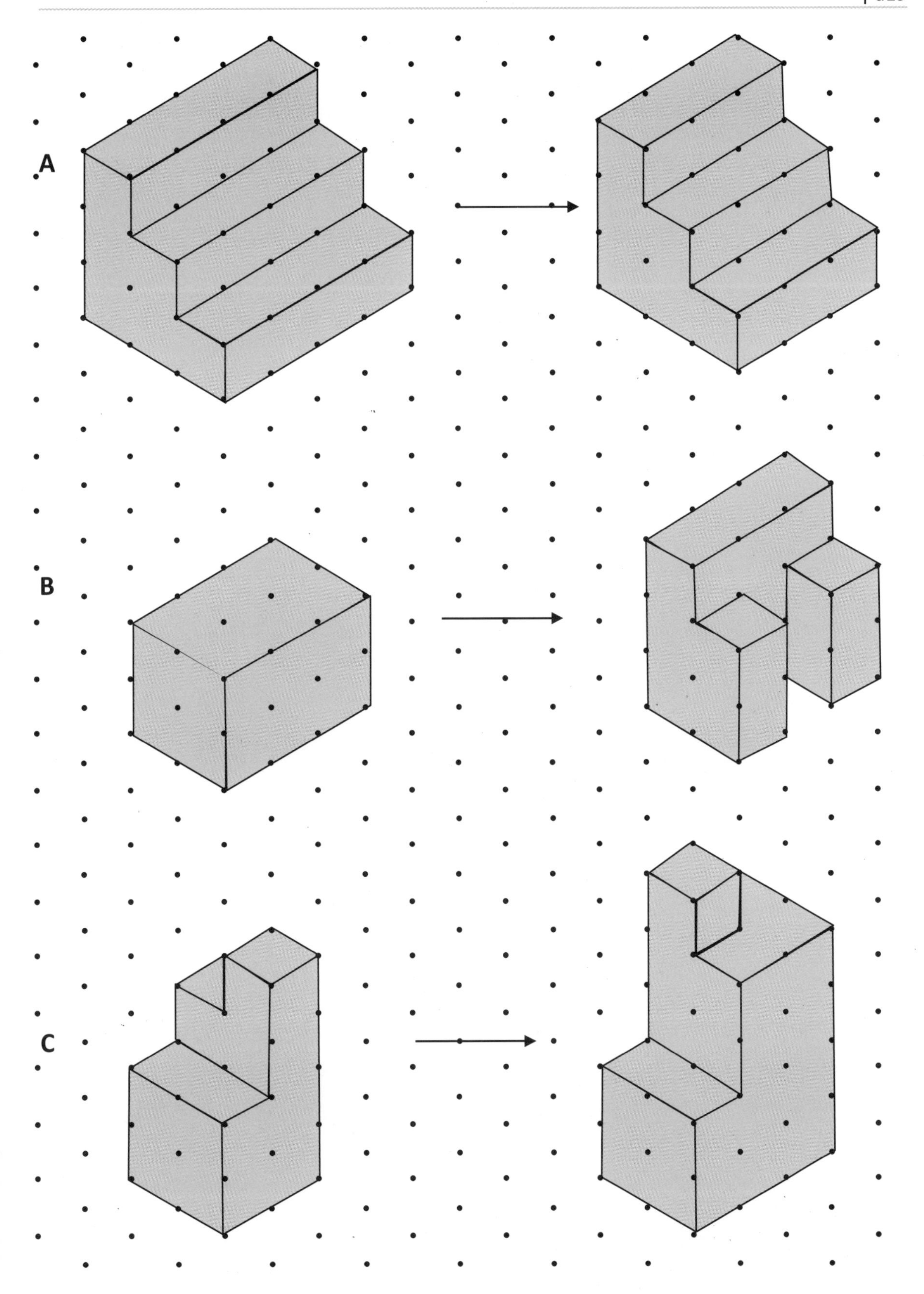
A
B
C

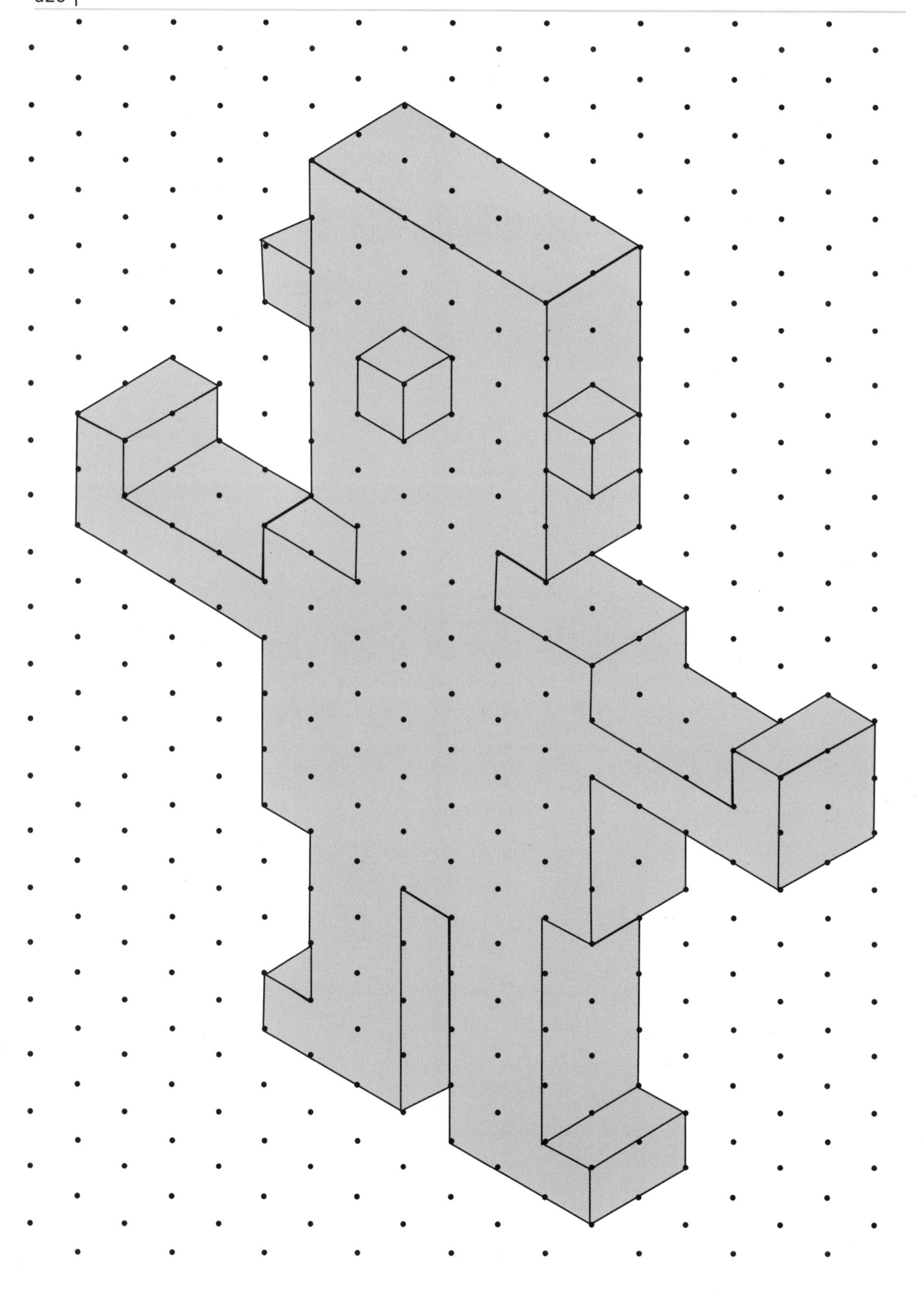

Blank page